Stefan Sperlich

Regularity and Integration Theory for a Class of Stochastic Processes

Stefan Sperlich

Regularity and Integration Theory for a Class of Stochastic Processes

Applications to Parabolic Problems

Südwestdeutscher Verlag für Hochschulschriften

Impressum / Imprint
Bibliografische Information der Deutschen Nationalbibliothek: Die Deutsche Nationalbibliothek verzeichnet diese Publikation in der Deutschen Nationalbibliografie; detaillierte bibliografische Daten sind im Internet über http://dnb.d-nb.de abrufbar.
Alle in diesem Buch genannten Marken und Produktnamen unterliegen warenzeichen-, marken- oder patentrechtlichem Schutz bzw. sind Warenzeichen oder eingetragene Warenzeichen der jeweiligen Inhaber. Die Wiedergabe von Marken, Produktnamen, Gebrauchsnamen, Handelsnamen, Warenbezeichnungen u.s.w. in diesem Werk berechtigt auch ohne besondere Kennzeichnung nicht zu der Annahme, dass solche Namen im Sinne der Warenzeichen- und Markenschutzgesetzgebung als frei zu betrachten wären und daher von jedermann benutzt werden dürften.

Bibliographic information published by the Deutsche Nationalbibliothek: The Deutsche Nationalbibliothek lists this publication in the Deutsche Nationalbibliografie; detailed bibliographic data are available in the Internet at http://dnb.d-nb.de.
Any brand names and product names mentioned in this book are subject to trademark, brand or patent protection and are trademarks or registered trademarks of their respective holders. The use of brand names, product names, common names, trade names, product descriptions etc. even without a particular marking in this works is in no way to be construed to mean that such names may be regarded as unrestricted in respect of trademark and brand protection legislation and could thus be used by anyone.

Coverbild / Cover image: www.ingimage.com

Verlag / Publisher:
Südwestdeutscher Verlag für Hochschulschriften
ist ein Imprint der / is a trademark of
AV Akademikerverlag GmbH & Co. KG
Heinrich-Böcking-Str. 6-8, 66121 Saarbrücken, Deutschland / Germany
Email: info@svh-verlag.de

Herstellung: siehe letzte Seite /
Printed at: see last page
ISBN: 978-3-8381-3595-3

Zugl. / Approved by: Halle (Saale), MLU, Diss., 2009

Copyright © 2012 AV Akademikerverlag GmbH & Co. KG
Alle Rechte vorbehalten. / All rights reserved. Saarbrücken 2012

Contents

Introduction   **5**

1 Foundations   **17**

    1.1 Fractional differintegration 18

    1.2 Function spaces 19

       1.2.1 Sequence spaces 19

       1.2.2 Spaces of nuclear and Hilbert-Schmidt operators 19

       1.2.3 Spaces of continuous, differentiable and Hölder functions 20

       1.2.4 Lebesgue spaces 21

       1.2.5 Spaces of random variables 22

       1.2.6 Bessel potential spaces, Besov spaces, Sobolev-Slobodeckij spaces 23

       1.2.7 Weighted spaces 25

    1.3 Evolutionary integral equations 27

2 Processes with stationary increments   **31**

    2.1 Definitions and Properties 32

    2.2 Regularity 37

- 2.3 Noise 48
- 2.4 Deterministic multipliers 49
- 2.5 Stochastic integration 55
  - 2.5.1 The real-valued case 55
  - 2.5.2 The vector-valued case 63
- 2.6 Examples 69
  - 2.6.1 Centered Poisson processes 69
  - 2.6.2 Fractional Brownian motions 72
  - 2.6.3 Fractional Riesz-Bessel motions 79

3 Parabolic Volterra equations — 87

- 3.1 Main results 87
- 3.2 Proof of the main results 92
  - 3.2.1 Proof of Theorem 3.1 95
  - 3.2.2 Proof of Theorem 3.5 96
- 3.3 The case $\alpha = 2$ 98

4 Anomalous diffusion — 99

- 4.1 Main results 101
- 4.2 Proof of the main results 104
  - 4.2.1 Weak solutions 104
  - 4.2.2 Proof of Theorem 4.2: Half-space setting. 106
  - 4.2.3 Spatial localization 108
  - 4.2.4 Proof of Theorem 4.2: Setting for domains. 112

CONTENTS 3

A Basic essentials 115

List of Figures 120

Bibliography 121

To Greta & Gustaf

Introduction

Random effects on mainly deterministic systems occur in many areas, for instance in flow mechanics or interest rate models. Usually Wiener processes are used to describe these random effects and a rich analytic toolbox was furnished for this Gaussian martingale process, on its forefront the stochastic integration calculi of Itô [56], Stratonovič [105] and Skorohod [102].

But Wiener type disturbances are Markov processes which means that they are no longer adequate if the data possess any chronological dependency. Indeed, studies have found that data in a large number of fields, including hydrology, geophysics, air pollution, image analysis, economics and finance display long-range dependence (e.g. Beran [15], Mandelbrot & Hudson [75], Peters [85]). To capture this phenomenon, Mandelbrot & van Ness [76] proposed in 1968 the concept of a fractional Brownian motion which, basically, is a probabilistic Gaussian process indexed by a parameter $H \in (0, 1)$. This parameter was named after the hydrologist Hurst who, together with some collaborators, demonstrated in the pioneering work [55] that this approach is appropriate to describe statistic time series in a hydrologically framework. Formally, a fractional Brownian motion is the convolution of Wiener increments with a power-law kernel. One of the advantages is, that one is able to control the stochastic influence by varying the parameter H in the interval $(0, 1)$. With the selection $H = \frac{1}{2}$ a fractional Brownian motion becomes a Wiener process which behaves chaotically since its increments are uncorrelated. Otherwise the increments of a fractional Brownian motion are negatively (if $H < \frac{1}{2}$), respectively positively (if $H > \frac{1}{2}$) correlated and in the last case this process exhibits long-range dependence, that is a certain memory feature, which is characterized by a spectral density of the incremental process having a singularity of some fractional order at frequency zero. Long-range dependence effects appear naturally in many situations, for example, when describing (cp. Shiryaev [101])

- The widths of consecutive annual rings of a tree.
- The temperature at a specific place as a function of time.
- The level of water in a river as a function of time.
- The characters of solar activity as a function of time.
- The values of the log returns of a stock.

Except for the Wiener case, a fractional Brownian motion is neither a semi-martingale nor Markovian and therefore extensive consequences can be observed if one simply modifies a stochastic model with replacing a Wiener process by a fractional Brownian motion. For instance in mathematical finance, Wiener processes are widely used to describe the movement of share prices (e.g. Prüss et al. [91]), but it is empirically demonstrated to be incorrect in a number of ways. As already mentioned, a fractional Brownian motion is in general not a semi-martingale, so particularly there cannot be a martingale measure (except for the case $H = \frac{1}{2}$), which by general results (e.g. Rogers [92], Cheridito [24]) means that there must be arbitrage. But this case is excluded by assumption in the common models. Nevertheless, fractional Brownian motions are of great interest in financial modeling (e.g. Elliott & van der Hoek [41], Hu [54], Necula [79], Jumarie [58], Liu & Yang [70, 71], Øksendal [83]), to say it with the words of Esko Valkeila: "As we all know, fractional Brownian motions cannot be used in finance, because it produces arbitrage. But as we also know, boys like to do forbidden things." As a consequence financial mathematicians tend to enlarge the common models with transaction costs and it was shown that in this richer framework fractional Brownian motions do no longer necessarily produce arbitrage (e.g. Guasoni [50]). In addition, recent studies detected a few more ways to exclude arbitrage (e.g.

Bender et al. [14]). However, the Wiener toolbox was not applicable for the theory around fractional Brownian motions, which made it necessary to establish a fully new stochastic calculus. This was done by many authors, among them Mandelbrot & van Ness [76], Lin [69], Dai & Heyde [28], Decreusefond & Üstünel [32, 33], Norros et al. [80], Duncan et al. [35, 37, 36], Alòs et al. [2, 1, 3], Pipiras & Taqqu [86], Krvavych & Mishura [66], Coutin et al. [26], Decreusefond [30, 31], Tudor [107], Lakhel et al. [68], Bender [13], Carmona et al. [23], Nualart [81, 82], Biagini et al. [19, 18, 17], Gradinaru et al. [49], Tudor [108], Jolis [57], Elliott & van der Hoek [42], and the progress is still going on.

In addition to long-range dependence, it has been found that many processes in finance (e.g. Bhansali et al. [16]) and 2-D turbulence in particular exhibit a high degree of intermittency, that is the clustering of extreme values at high frequencies of a certain order, so for instance a multiplicative cascade process (e.g. Davis et al. [29]). Intermittency can be loosely described as the characteristic of a dynamic system, whose substantially regular behavior is interspersed by infrequent and compendious chaotic phases. Intermittency effects occur, for example, when describing (cp. Shiryaev [101])

- Financial turbulence, e.g. the empirical volatility of a stock.
- The prices of electricity in a liberated electricity market.

In 1999, Anh et al. proposed in [6] a two parameter process called fractional Riesz-Bessel motion, which may exhibit both, long-range dependence and second-order intermittency. In other words, the presence of a fractional Riesz-Bessel motion affords a possibility to study both effects simultaneously. This study was undertaken by Anh et al. [8]. While a Wiener process is a special case of a fractional Brownian motion, the last is on the other hand a special case of a fractional Riesz-Bessel motion. However, again people were facing the problem, that the stochastic calculus for fractional Brownian motions did not fit to fractional Riesz-Bessel motions, since in the past the calculi were tailor-made for each process.

Therefore it would be desirable to have a rigorous stochastic analysis for a satisfactory large class of stochastic processes, say for stochastic processes with stationary increments and spectral density. This is the major aim of the present thesis. Once provided, we will present applications to parabolic problems arising frequently in models concerning linear viscoelastic material behavior and fractional diffusion.

This thesis is structured as follows. In Chapter 1 we explain some mathematical notations and function spaces and we introduce briefly the fundamentals of evolutionary integral equations, which are widely taken from the monograph of Prüss [88].

In Chapter 2 we define real-valued and also vector-valued processes with stationary increments and prove regularity results for certain classes of those motions. Precisely, we consider two classes of processes characterized by Hypotheses (ϕ) and (ϕ_0) (see page 37). A process X with stationary increments and spectral density ϕ satisfies Hypothesis (ϕ), if there is a number $\gamma \in (1,3)$, such that $|\lambda|^\gamma \phi(\lambda)$ is bounded on the real line $\mathbb{R}$. On the other hand, X is due to Hypothesis (ϕ_0), if a number $\gamma_0 \in (1,3)$ exists, so that $0 < |\lambda|^{\gamma_0} \phi(\lambda) < \infty$ in a certain neighborhood of zero and if the spectral density ϕ satisfies a growth condition (this condition will determine a number $\theta \geq 0$). As an additional benefit, the numbers γ, γ_0 and θ provide information whether the process X may exhibit long-range dependence or intermittency (see Remark 2.5). The most employed result of the present thesis is formulated in Theorem 2.11 and reads as

Theorem. *The following are true.*

(i) Let X be subject to Hypothesis (ϕ). Then there is a constant $c_\phi > 0$, such that the estimate

$$\mathbb{E}[X(\tau)]^2 \leq c_\phi |\tau|^{\gamma-1}$$

holds for all $\tau \in \mathbb{R}$. Moreover, we have equality if $|\lambda|^\gamma \phi(\lambda)$ is constant.

(ii) Let X be subject to Hypothesis (ϕ_0). Then there is a number $c_{\phi_0} > 0$, such that the estimate

$$\mathbb{E}[X(\tau)]^2 \geq c_{\phi_0} \cdot \min\{|\tau|^{\gamma_0-1+\theta}, |\tau|^{\gamma_0-1}\}$$

holds for all $\tau \in \mathbb{R}$.

In particular, we will be able to prove that in case X is centered and satisfies Hypotheses (ϕ) and (ϕ_0), the variance $\mathrm{Var}[X(t)]$ takes values in the shaded regions of Figure 1.

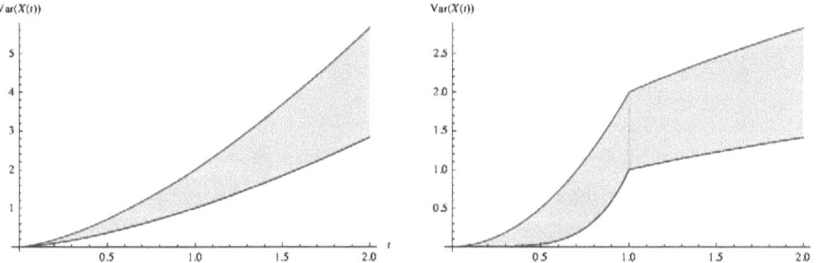

Figure 1. Idealized regions for the values of $\mathrm{Var}[X(t)]$, where X is centered and due to (ϕ) and (ϕ_0) with $\theta = 0$ (left) and $\theta > 0$ (right), respectively.

Figure 1 is idealized in the sense that the shaded regions might be thinner, thicker, steeper or shallower. This depends on the concrete values of the parameters γ, γ_0, θ and the constants c_ϕ and c_{ϕ_0}. The occurrence at time $t = 1$ when $\theta > 0$ is here exaggerated for the sake of clarity but, however, is significant and strongly connected to the appearance of intermittency. Regarding long-range dependence the result is also useful, since as a rule of thumb the process X may display this property only if the function $\mathrm{Var}[X(t)]$ increases with super-linear order. These estimates will be employed frequently in Section 2.2 to obtain multitude regularity results, so for instance results in the pathwise sense captured by Theorem 2.18

Theorem. *The following are true.*

(i) *Let X be subject to Hypothesis (ϕ). If $\gamma > 2$, then X is mean-square continuous and has continuous paths almost surely. Moreover, with probability 1, the trajectories of X are locally Hölder-continuous of any order strictly less than $\frac{\gamma-2}{2}$.*

(ii) *Let X be a centered Gaussian process subject to Hypothesis (ϕ). Then X is mean-square continuous and has continuous paths almost surely. Moreover, with probability 1, the trajectories of X are locally Hölder-continuous of any order strictly less than $\frac{\gamma-1}{2}$.*

(iii) *Let X be subject to Hypothesis (ϕ_0). If $\theta < 3 - \gamma_0$ then X is almost surely nowhere mean-square differentiable.*

Due to available $L_p(\Omega)$-estimates, deduced from the Kahane-Khintchine inequality (cp. Theorem A.3), the results in Gaussian case perform consistently better. Regarding temporal regularity in the $L_p(\Omega)$-sense, Theorem 2.21 yields

Theorem. *Let $T > 0$, $J = [0,T]$, $p \in (0,\infty)$ and $0 < \sigma < 1$.*

(i) *Suppose X satisfies Hypothesis (ϕ). If $2\sigma < \gamma - 1$, then $X \in {}_0W_2^\sigma(J; L_2(\Omega))$.*

(ii) *Suppose X satisfies Hypothesis (ϕ_0). If $2\sigma \geq \gamma_0 - 1 + \theta$, then $X \notin {}_0W_2^\sigma(J; L_2(\Omega))$.*

(iii) *Suppose X is a centered Gaussian process subject to Hypothesis (ϕ) and let $2 \leq q < \infty$. If $2\sigma < \gamma - 1$, then $X \in {}_0W_p^\sigma(J; L_q(\Omega))$.*

(iv) *Suppose X is a centered Gaussian process subject to Hypothesis (ϕ_0) and let $1 < q \leq 2$. If $2\sigma \geq \gamma_0 - 1 + \theta$, then $X \notin {}_0W_p^\sigma(J; L_q(\Omega))$.*

Then, in Section 2.4, we spare some time with deterministic multipliers and study the question: Given a sequence of mutually independent processes $(X_n)_{n \in \mathbb{N}}$, what are necessary and sufficient conditions on the multiplier $b := (b_n)_{n \in \mathbb{N}}$, such that the function

$$\zeta(t,x,\omega) := \sum_{n=1}^{\infty} b_n(t,x) X_n(t,\omega)$$

affiliates to a given regularity class? Having answered this question we accomplish to stochastic integrals of deterministic integrants with respect to a process X with stationary increments and a spectral density ϕ. The result in the real-valued case, stated in Theorem 2.30, holds independently of Hypotheses (ϕ) or (ϕ_0) and allocates the isometry of Itô-type

$$\mathbb{E}\left[\left(\int_{\mathbb{R}} f(\tau)\mathrm{d}X(\tau)\right)\left(\int_{\mathbb{R}} g(\tau)\mathrm{d}X(\tau)\right)\right] = \int_{\mathbb{R}} (\mathcal{F}f)(\lambda)\overline{(\mathcal{F}g)(\lambda)}\lambda^2\phi(\lambda)\mathrm{d}\lambda,$$

which is true for all functions (or distributions) f and g for which the right hand-side is meaningful and finite. The innovative impact of this isometry is that we do not have to impose a probabilistic distribution of the motion X. It holds true for any stochastic process with stationary increments whose spectral density exists, so for instance it is valid for centered Lévy processes, fractional Brownian motions and fractional Riesz-Bessel motions for any thinkable choice of parameters (see Section 2.6 for an elaborate treatise of this examples). As a matter of course, we will present similar results for the vector-valued case (see Theorems 2.32 & 2.33).

The remaining part of this thesis is devoted to parabolic problems with perturbations involving processes under consideration. In the focus of Chapter 3, there are two types of parabolic Volterra equations. Letting $\mathcal{H}$ be a separable Hilbert space, we first consider the problem

$$u + (b * Au) = Q^{1/2}\mathcal{X} \tag{VE1}$$

on the half-line $\mathbb{R}_+$, where $Q^{1/2}\mathcal{X}$ denotes a system independent $\mathcal{H}$-valued process of spectral type ϕ. With system independence we mean that the eigensystems of the operators Q and A do not have to coincide. Here we choose the natural framework which is typically in the theory of linear viscoelastic material behavior, that is the operator $-A$ behaves as an elliptic differential operator like the Laplacian, the elasticity operator, or the Stokes operator, together with appropriate boundary conditions (cp. Prüss [90, Section 5]). The kernel b is assumed to be the antiderivative of a 3-monotone scalar function, think of the material functions of Newtonian fluids, Maxwell fluids or of power type materials. The explicit assumptions on the operator A, the kernel b and the process $Q^{1/2}\mathcal{X}$ are formulated in Hypotheses *(A)*, *(b)* (see page 88f) and *(X_ϕ)* (see page 68), respectively. With the aid of the most important property, that is the self-adjointness of the operator A, we derive sharp estimates such that the mild solution's trajectories are Hölder-continuous in time up to a certain order. The proven results are consistent with those of Clément et al. [25], where a differentiated version of problem (VE1) with an A-synchronized white noise disturbance was studied. The terminus of an A-synchronized perturbation links to coinciding eigensystems of the operators A and Q. Unless the synchronized case is interesting from a mathematical viewpoint, it seems to be too restrictive for applications, because this case, roughly speaking, corresponds to disturbances acting solely on the system's eigenfrequency. However, we will show that the mild solution's properties in terms of existence, uniqueness and pathwise regularity do not differ in both cases. Denoting by $\mathscr{L}_1$ the space of nuclear operators (see Section 1.2.2) and setting

$$\rho := \frac{2}{\pi} \sup\{|\arg \widehat{b}(\lambda)| : \operatorname{Re} \lambda > 0\},$$

where $\widehat{b}$ means the Laplace transform of b, the main result concerning problem (VE1) is stated in Theorem 3.1 and reads as

Theorem. *Let Hypotheses (A), (b) and (X_ϕ) are valid.*

(i) *If $QA^{\frac{1-\gamma}{\rho}} \in \mathscr{L}_1(\mathcal{H})$, then the mild solution u of (VE1) exists and is mean-square continuous on $\mathbb{R}_+$. Moreover, the trajectories of u are continuous on the half-line $\mathbb{R}_+$ almost surely.*

(ii) *If in addition, there is $\theta \in (0, \frac{\gamma-1}{2})$ such that $QA^{\frac{1-\gamma}{\rho}+\frac{2\theta}{\rho}} \in \mathscr{L}_1(\mathcal{H})$, then the trajectories of u are locally Hölder-continuous of any order strictly less then θ almost surely.*

We then take up a different view point to Volterra equations with noise, i.e. we study the problem

$$u + (g_\alpha * Au) = (g_\beta * Q^{1/2}\dot{\mathcal{X}}) \quad \text{(VE2)}$$

on the half-ray $\mathbb{R}_+$, where g_κ denotes the Riemann-Liouville kernel of fractional integration; see (1.4). We then obtain in virtue of Theorem 3.5

Theorem. Assume Hypotheses (A) and (X_ϕ) are valid and let $\alpha \in (0,2)$, $\beta > 0$, $\theta \in [0,1]$, such that $\beta \in (\frac{3-\gamma}{2} + \theta, \frac{3-\gamma}{2} + \theta + \alpha)$.

(i) If $QA^{\frac{3-2\beta-\gamma}{\alpha}} \in \mathscr{L}_1(\mathcal{H})$ then the mild solution u of (VE2) exists and is mean-square continuous on $\mathbb{R}_+$. Moreover, the trajectories of u are almost surely continuous on $\mathbb{R}_+$.

(ii) If $QA^{\frac{3-2\beta-\gamma}{\alpha}+\frac{2\theta}{\alpha}} \in \mathscr{L}_1(\mathcal{H})$ then the trajectories of u are locally Hölder-continuous of any order strictly less then θ almost sure.

Similar results for the special cases where $\mathcal{X}$ is modeled to be a A-synchronized vector-valued Wiener process or a vector-valued fractional Brownian motion were obtained by Clément et al. [25], Bonaccorsi [20] and Sp. & Wilke [104]. Results for a system independent vector-valued fractional Brownian motion are available by Sp. [103]. However, all those cases are completely covered by our approach.

Finally, in Chapter 4 we put our attention to problems of anomalous diffusion, that is

$$\begin{cases} \partial_t^\alpha u - \Delta u = 0, \\ \mathcal{D}u|_{\partial G} = \psi, \\ u|_{t=0} = 0, \end{cases} \quad \text{(AD)}$$

where $\alpha \in (0,2)$, $G \subset \mathbb{R}^N$ is a domain with a somehow smooth boundary and ψ is a stochastic boundary perturbation modeled as

$$\psi(t,x\omega) = \sum_{n=1}^\infty b_n(t,x) X_n(t,\omega),$$

where $b := (b_n)_{n \in \mathbb{N}}$ is a sequence of appropriate scalar functions and $(X_n)_{n \in \mathbb{N}}$ is a sequence of mutually independent processes of a certain type. The symbol $\mathcal{D}$ means the identity mapping or the derivative in normal direction, selectively. So the formulation of system (AD) covers both, the corresponding Dirichlet and the Neumann problem.

Fractional diffusion equations were introduced to describe physical phenomena such as diffusion on porous media with fractal geometry, kinematics in viscoelastic media, relaxation processes in complex systems (including viscoelastic materials, glassy materials, synthetic polymers, biopolymers), propagation of seismic waves, anomalous diffusion and turbulence (see Caputo [22], Glöckle & Nonnenmacher [46], Mainardi & Paradisi [74], Saichev & Zaslavsky [98], Mainardi & Gorenflo [73, 48], Kobelev et al. [64, 63, 62], Hilfer [53] and the references therein). These equations are obtained from the classical diffusion equation by replacing the first or second order derivative by a fractional derivative (see Section 1.1 and also Oldham & Spanier [84], Samko et al. [99], Miller & Ross [77], Gorenflo & Mainardi [47], Džrbašjan & Nersesjan [40], Podlubny [87], Butzer & Westphal [21] for different types of fractional derivatives, fractional integrals or fractional operators in general and their properties). Even in finance, the fractional diffusion equations are of importance. So for instance in the theory of tick-by-tick dynamics in financial markets (cf. Scalas et al. [100]), where the general scaling form can be obtained as the solution of a certain fractional diffusion equation. For brevity we designate

$$U_{\delta,\gamma} := {}_0W_{2,\frac{\gamma-1}{2}}^{\frac{\alpha\delta}{4}}(J; L_2(\partial G; \ell_2)) \cap L_{2,\frac{\gamma-1}{2}}\left(J; {}_0W_2^{\frac{\delta}{2}}(\partial G; \ell_2)\right), \quad \delta \geq 0,$$

$$U_{\delta,\gamma}^0 := {}_0W_2^{\frac{\alpha\delta}{4}}(J; L_2(\partial G; \ell_2)) \cap L_{2,\frac{\gamma-1}{2}}\left(J; {}_0W_2^{\frac{\delta}{2}}(\partial G; \ell_2)\right), \quad \delta \geq 0.$$

and also

$$Z_\delta := {}_0W_2^{\frac{\alpha\delta}{4}}(J; L_2(G; L_2(\Omega))) \cap L_2\left(J; {}_0W_2^{\min\{\frac{\delta}{2};2\}}(G; L_2(\Omega))\right), \quad \delta \geq 0.$$

Summarizing the explicit assumptions on the disturbance ψ in Hypotheses (ψ) and (ψ_0) (see page 99), our main result is stated in Theorem 4.2 and reads as

Theorem. *Assume Hypothesis (ψ) holds. Let $0 \leq \nu < \frac{2(\gamma-1)}{\alpha}$ and in case $G \neq \mathbb{R}^N_+$ let $\nu \in [0, \frac{2(\gamma-1)}{\alpha}) \cap [0, 4)$. Then the following hold if $b \in U^0_{\nu,\gamma}$.*

(i) *The Dirichlet problem (AD), i.e. $\mathcal{D} = I$, admits a unique solution u in the regularity class $Z_{\nu+1}$. If, in addition, $\nu \leq 3$ and Hypothesis (ψ_0) is valid, then membership of b to the class $U_{\nu,\gamma}$ is necessary and sufficient.*

(ii) *The Neumann problem (AD), i.e. $\mathcal{D} = \partial_\nu$, admits a unique solution u in the regularity class $Z_{\nu+3}$. If, in addition, $\nu \leq 1$ and Hypothesis (ψ_0) is valid, then membership of b to the class $U_{\nu,\gamma}$ is necessary and sufficient.*

Here the number $1 < \gamma < 3$ is determined by Hypothesis (ψ), which is strongly connected to Hypothesis (ϕ) introduced earlier in Section 2.2. There are already several results concerning stochastic boundary value problems (e.g. Rozanov & Sanso [94], Kijima & Suzuki [61], Rößler et al. [93]), but to the author's knowledge, results for the fractional diffusion equation with random boundary conditions are still rare.

Chapter 1

Foundations

In what follows let X and Y be Banach spaces and $\mathcal{H}$ be a separable Hilbert space. $J \subset [0, \infty)$ will usually mean a bounded or unbounded time interval. We endeavor to denote the norm in X with $\|\cdot\|_X$, but from time to time we may write $\|\cdot\|$ or $|\cdot|_X$ if it is conducive to brevity. An inner product will be denoted by $(\cdot|\cdot)$ and if there is any risk of confusion we will add a lower index to designate the affiliation to a certain inner product space.

By $\mathbb{N}, \mathbb{R}, \mathbb{C}$ we denote the sets of natural, real and complex numbers, respectively, and let further $\mathbb{R}_+ = [0, \infty)$, $\mathbb{C}_+ = \{\lambda \in \mathbb{C} : \operatorname{Re} \lambda > 0\}$, $\mathbb{N}_0 = \mathbb{N} \cup \{0\}$. The symbol $\mathcal{B}(X;Y)$ means the space of all bounded linear operators from X to Y and we write $\mathcal{B}(X) = \mathcal{B}(X;X)$ for short.

If A is an operator in X, $\mathcal{D}(A)$ and $\mathcal{R}(A)$ stand for domain and range of A, respectively, while $\rho(A)$, $\sigma(A)$ designate the resolvent set and the spectrum of A.

As usual we employ the star $*$ for the convolution of functions defined on the line $\mathbb{R}$

$$(f * g)(t) = \int_{-\infty}^{\infty} f(t - \tau) g(\tau) \mathrm{d}\tau, \qquad t \in \mathbb{R}, \tag{1.1}$$

and

$$(f * g)(t) = \int_{0}^{t} f(t - \tau) g(\tau) \mathrm{d}\tau, \qquad t \geq 0, \tag{1.2}$$

for f, g supported on the half-ray $\mathbb{R}_+$. Observe that (1.1) and (1.2) are equivalent for functions which vanish on $(-\infty, 0)$; therefore there will be no danger of confusion.

For $u \in L_{1,\mathrm{loc}}(\mathbb{R}_+; X)$ of exponential growth, i.e. $\int_0^\infty e^{-\omega t}|u(t)|\mathrm{d}t < \infty$ with some $\omega \in \mathbb{R}$, the Laplace transform of u is defined by

$$\widehat{u}(\lambda) = \int_0^\infty e^{-\lambda t} u(t) \mathrm{d}t, \quad \operatorname{Re}\lambda \geq \omega.$$

For $f \in L_1(\mathbb{R}; X)$, the Fourier transform of f is the function $\mathcal{F}f : \mathbb{R} \to X$ defined by

$$(\mathcal{F}f)(\xi) = \int_\mathbb{R} e^{-i\xi t} f(t) \mathrm{d}t.$$

Throughout this thesis we will denote by χ_M the characteristic function of the set M, that is $\chi_M(x) = 1$ if $x \in M$ and $\chi_M(x) = 0$ otherwise.

1.1 Fractional differintegration

The concept of differentiation and integration of noninteger order has a long history. Interest in this subject was evident almost as soon as the ideas of the classical calculus were known. Some of the earliest more or less systematic studies seem to have been made in the 18th and 19th century by Euler, Lagrange, Liouville, Riemann and Holmgren.

Within this thesis we make use of the notion of the (left-sided) fractional differintegral of order $\alpha \in (-2, 2)$ of a test-function ϕ by $\partial^\alpha \phi$ being defined as

$$(\partial^\alpha \phi)(t) := \frac{\mathrm{d}^2}{(\mathrm{d}t)^2} \int_{-\infty}^t g_{2-\alpha}(t-\tau)\phi(\tau)\mathrm{d}\tau, \quad t \in \mathbb{R}, \tag{1.3}$$

where

$$g_\kappa(t) = \frac{t^{\kappa-1}}{\Gamma(\kappa)}, \quad t \geq 0, \quad \kappa > 0 \tag{1.4}$$

denotes the Riemann-Liouville kernel. Note that g_κ is of subexponential growth, i.e.

$$\int_0^\infty e^{-\omega t}|g_\kappa(t)|\mathrm{d}t < \infty$$

1.2. FUNCTION SPACES

for arbitrary small $\omega > 0$. This means that the Laplace transform $\widehat{g}_\kappa$ of g_κ is well-defined, and we have

$$(\widehat{\partial^\alpha \phi})(\lambda) = \lambda^\alpha \widehat{\phi}(\lambda), \qquad \operatorname{Re}\lambda > 0$$

for all test-functions ϕ supported on $(0, \infty)$.

1.2 Function spaces

Aim of this section is to give meaning to function spaces of interest for the present thesis. Throughout this section X will be a Banach space, if not indicated otherwise.

1.2.1 Sequence spaces

By ℓ_p we denote the sequence space of real- or complex-valued sequences $a := (a_n)_{n \in \mathbb{N}}$

$$\ell_p = \left\{ a : \sum_{n=1}^\infty |a_n|^p < \infty \right\}, \qquad 1 \leq p < \infty,$$

equipped with the norm

$$\|a\|_p = \left[\sum_{n=1}^\infty |a_n|^p \right]^{\frac{1}{p}}.$$

It is well-known that $(\ell_p, \|\cdot\|_p)$ is a Banach space and a Hilbert space if and only if $p = 2$. The inner product in ℓ_2 then reads $(a|b)_2 = \sum_{n=1}^\infty a_n \overline{b_n}$. As a general reference towards sequence spaces we refer to Dunford & Schwartz [38, Chapter IV.2].

1.2.2 Spaces of nuclear and Hilbert-Schmidt operators

In what follows let $\mathcal{H}$ be a separable Hilbert space. The symbols $\mathscr{L}_1(\mathcal{H})$ and $\mathscr{L}_2(\mathcal{H})$ denote the spaces of nuclear operators and Hilbert-Schmidt operators on $\mathcal{H}$, respectively. Thereby a bounded operator T on $\mathcal{H}$ is called nuclear (that is $T \in \mathscr{L}_1(\mathcal{H})$) if there are sequences $(x_n^*) \subset \mathcal{H}^*$ and $(y_n) \subset \mathcal{H}$ with $\sum_{n=1}^\infty \|x_n^*\|\|y_n\| < \infty$ so that

$$Tx = \sum_{n=1}^\infty x_n^*(x) y_n$$

holds for all $x \in \mathcal{H}$. On the other hand a bounded operator T on $\mathcal{H}$ is said to be a Hilbert-Schmidt operator (meaning $T \in \mathscr{L}_2(\mathcal{H})$), if there is an orthonormal basis $(e_n) \subset \mathcal{H}$, so that

$$\sum_{n=1}^\infty \|Te_n\|^2 < \infty.$$

If this is true for one orthonormal basis, it is true for any other orthonormal basis of $\mathcal{H}$. We have

$$\mathscr{L}_1(\mathcal{H}) \hookrightarrow \mathscr{L}_2(\mathcal{H}) \hookrightarrow \mathcal{B}(\mathcal{H}).$$

For an elaborate treatise to these spaces we refer to Dunford & Schwartz [39, Chapter XI.6] and Da Prato & Zabczyk [27, Appendix C]. In case the operator $T : \mathcal{H} \to \mathcal{H}$ is self-adjoint with eigenvalues $\lambda = (\lambda_n)_{n \in \mathbb{N}}$, the norms in these spaces can be written as

$$\|T\|_{\mathscr{L}_1(\mathcal{H})} = \|\lambda\|_{\ell_1},$$
$$\|T\|_{\mathscr{L}_2(\mathcal{H})} = \|\lambda\|_{\ell_2}.$$

For nuclear operators T on $\mathcal{H}$ one can define the trace of T by means of

$$\mathrm{Tr}[T] = \sum_{n=1}^\infty (Tg_n \mid g_n)_{\mathcal{H}},$$

where $(g_n)_{n \in \mathbb{N}}$ is an arbitrary orthonormal basis in $\mathcal{H}$. Due to this property nuclear operators are also called operators of trace class. One can show, that $|\mathrm{Tr}[T]| \leq \|T\|_{\mathscr{L}_1(\mathcal{H})}$ holds for every $T \in \mathscr{L}_1(\mathcal{H})$ and, moreover, that $\mathrm{Tr}[T] = \|T\|_{\mathscr{L}_1(\mathcal{H})}$ if T is positive semi-definite.

1.2.3 Spaces of continuous, differentiable and Hölder functions

Let $U \subset X$ be open, then $C(U; Y)$ and $C_b(U; Y)$ stand for the spaces of continuous resp. bounded continuous functions $f : X \to Y$. Those spaces equipped with the

1.2. FUNCTION SPACES

norm
$$\|f\|_\infty = \sup\{|f(x)|_Y : x \in U\}$$
are Banach spaces. For $m \in \mathbb{N}$, the symbol $C_b^m(U;Y)$ means the space of all m-times continuously differentiable functions $f : U \to Y$ with norm
$$\|f\|_m = \sum_{|\alpha|\le m} \|D^\alpha f\|_\infty.$$
The space $(C_b^m(U;X), \|\cdot\|_m)$ is a Banach space. With $C^\infty(U;Y)$ we denote the function space containing all smooth functions, meaning all functions which are infinitely often differentiable.

Further, if $\alpha \in (0,1)$, then $C_b^\alpha(U;Y)$ designates the space of all Hölder-continuous functions $f : U \to Y$ of order α normed by
$$\|f\|_\alpha = \|f\|_\infty + \sup\left\{\frac{|f(x)-f(y)|_Y}{|x-y|_X^\alpha} : x,y \in U, x \ne y\right\}.$$
Every Hölder-continuous function is uniformly continuous. If $\alpha > 1$ is not an integer, we set $\alpha = [\alpha] + \{\alpha\}$, where $[\alpha]$ is an integer and $0 < \{\alpha\} < 1$. Then $C_b^\alpha(U;Y)$ means the space of all functions $f : U \to Y$, whose $[\alpha]$-th derivative exists and belongs to $C_b^{\{\alpha\}}(U;Y)$.

1.2.4 Lebesgue spaces

Let $D \subset \mathbb{R}^N$ be a Lebesgue-measurable set and $1 \le p < \infty$. Then $L_p(D;X)$ denotes the space of all (equivalence classes of) Bochner-measurable functions $f : D \to X$ so that
$$\|f\|_p := \left[\int_D |f(x)|_X^p \, dx\right]^{\frac{1}{p}} < \infty.$$
$L_p(D;X)$ is a Banach space when normed by $\|\cdot\|_p$ and a Hilbert space if and only if $p = 2$ and X is a Hilbert space. In this case we have the L_2-inner product
$$(f \mid g)_{L_2(D;X)} = \int_D (f(x) \mid g(x))_X \, dx.$$

Similarly, $L_\infty(D; X)$ stands for the space of (equivalence classes of) Bochner-measurable functions $f : D \to X$, with norm

$$\|f\|_\infty := \operatorname{ess\,sup} \{|f(x)|_X : x \in D\}.$$

With this norm, $L_\infty(D; X)$ is a Banach space. The subscript loc assigned to any of the above function spaces means the membership to the corresponding space when restricted to compact subsets of its domain.

1.2.5 Spaces of random variables

Let $(\Omega, \mathcal{F}, \mathbb{P})$ be a probability space and $\mathcal{H}$ be a separable Hilbert space. A random variable $\xi : \Omega \to \mathcal{H}$ is said to be (Bochner-)integrable if

$$\int_\Omega \|\xi(\omega)\|_\mathcal{H} \mathbb{P}(\mathrm{d}\omega) < \infty$$

and we define the expectation operator $\mathbb{E}$ as the integral

$$\mathbb{E}[\xi] := \int_\Omega \xi \,\mathrm{d}\mathbb{P}.$$

The symbol $L_1(\Omega, \mathcal{F}, \mathbb{P}; \mathcal{H})$ denotes the set of (all equivalence classes of) $\mathcal{H}$-valued random variables. Equipped with the norm

$$\|\xi\|_{L_1(\Omega)} = \mathbb{E}[\|\xi\|_\mathcal{H}]$$

the space $L_1(\Omega, \mathcal{F}, \mathbb{P}; \mathcal{H})$ is a Banach space. In a similar way one can define $L_p(\Omega, \mathcal{F}, \mathbb{P}; \mathcal{H})$, for arbitrary $p > 1$ with norms

$$\|\xi\|_{L_p(\Omega)} = (\mathbb{E}[\|\xi\|_\mathcal{H}^p])^{1/p}, \qquad 1 < p < \infty,$$

and

$$\|\xi\|_{L_\infty(\Omega)} = \operatorname{ess\,sup} \{\|\xi\|_\mathcal{H} : \omega \in \Omega\}.$$

If there is no risk of confusion we will write for short $L_p(\Omega)$ instead of $L_p(\Omega, \mathcal{F}, \mathbb{P}; \mathcal{H})$. Moreover, for arbitrary elements $x, y \in \mathcal{H}$ we denote by $x \otimes y$ the linear operator defined by

$$(x \otimes y)h = x(y \mid h)_\mathcal{H}, \qquad h \in \mathcal{H}.$$

1.2. FUNCTION SPACES

For ξ, η belonging to $L_2(\Omega, \mathcal{F}, \mathbb{P}; \mathcal{H})$ we follow Da Prato & Zabczyk [27] and introduce the covariance operator of ξ and of (ξ, η) by the formulae

$$\mathrm{Cov}(\xi) := \mathbb{E}[(\xi - \mathbb{E}[\xi]) \otimes (\xi - \mathbb{E}[\xi])],$$
$$\mathrm{Cov}(\xi, \eta) := \mathbb{E}[(\xi - \mathbb{E}[\xi]) \otimes (\eta - \mathbb{E}[\eta])].$$

Note that $\mathrm{Cov}(\xi)$ is a symmetric, positive, and nuclear operator with

$$\mathrm{Tr}[\mathrm{Cov}(\xi)] = \mathbb{E}\left[\|\xi - \mathbb{E}[\xi]\|_\mathcal{H}^2\right] =: \mathrm{Var}(\xi).$$

1.2.6 Bessel potential spaces, Besov spaces, Sobolev-Slobodeckij spaces

For an open subset $D \subset \mathbb{R}^N$, $\mathrm{H}_p^m(D; X)$ with $m \in \mathbb{N}$ denotes the classical Sobolev space, that is the space of all functions $f : D \to X$ having distributional derivatives $\partial^\alpha f \in L_p(D; X)$ of order $0 \leq |\alpha| \leq m$. For $1 \leq p < \infty$ the norm in $\mathrm{H}_p^m(D; X)$ is given by

$$\|f\|_{\mathrm{H}_p^m(D;X)} := \left[\sum_{|\alpha| \leq m} \|\partial^\alpha f\|_p^p\right]^{\frac{1}{p}}.$$

Further, for $0 < s < 1$, we define the Bessel potential spaces $\mathrm{H}_p^{sm}(D; X)$, by means of complex interpolation via

$$\mathrm{H}_p^{sm}(D; X) := \left[L_p(D; X); \mathrm{H}_p^m(D; X)\right]_s.$$

We will from time to time also use the Besov spaces $B_{pq}^{sm}(D; X)$ which can be defined via real interpolation by

$$B_{pq}^{sm}(D; X) := \left(L_p(D; X); \mathrm{H}_p^m(D; X)\right)_{s,q}, \quad 0 < s < 1, \ 1 \leq p < \infty, \ 1 \leq q \leq \infty.$$

Recall that $B_{pp}^s(D; X) = W_p^s(D; X)$, provided that $s \notin \mathbb{N}$, where $W_p^s(D; X)$ denotes the Sobolev-Slobodeckij space. For a general definition of these spaces we refer to Triebel [106] or Runst & Sickel [97]. Note further, that in case $p = 2$ and X is a Hilbert or UMD space (see e.g. Amann [4] for the definition and properties of UMD spaces) we have

$$\mathrm{H}_2^s(D;X) = W_2^s(D;X), \qquad s \geq 0.$$

With $s = [s] + \{s\}$, where $[s]$ is an integer and $0 < \{s\} < 1$, the intrinsic norm in $W_p^s(\mathbb{R}^N; X)$ can be written as

$$\|f\|_{W_p^s(\mathbb{R}^N;X)} = \|f\|_{W_p^{[s]}(\mathbb{R}^N;X)} + \sum_{|\alpha|=[s]} \left(\int_{\mathbb{R}^N} \int_{\mathbb{R}^N} \frac{|\partial^\alpha f(x) - \partial^\alpha f(y)|_X^p}{|x-y|^{N+p\{s\}}} \mathrm{d}x \mathrm{d}y \right)^{\frac{1}{p}}, \qquad s > 0. \tag{1.5}$$

Note, that the second term from the right hand-side of (1.5) defines a semi-norm in $W_p^s(\mathbb{R}^N; X)$, which will be abbreviated by $[f]_{W_p^s(\mathbb{R}^N;X)}$ if necessary.

Then, by $\mathcal{S}^*(\mathbb{R}^N)$ we denote the space of tempered distributions, the topological dual of the Schwartz space $\mathcal{S}(\mathbb{R}^N)$ and recall that for $1 \leq p \leq \infty$ and $g \in L_p(\mathbb{R}^N)$ the operator

$$T_g(\phi) = \int_{\mathbb{R}^N} g(x)\phi(x)\mathrm{d}x$$

defines a tempered distribution, i.e. $T_g \in \mathcal{S}^*(\mathbb{R}^N)$, so $L_p \subset \mathcal{S}^*(\mathbb{R}^N)$ for all $1 \leq p \leq \infty$. Recall further, that for $f \in \mathcal{S}^*(\mathbb{R}^N)$ the Fourier transform $\mathcal{F}f$ is well-defined and given by

$$(\mathcal{F}f)(\phi) = f(\mathcal{F}\phi) \qquad \text{for all} \qquad \phi \in \mathcal{S}(\mathbb{R}^N).$$

Since $\mathcal{F} : \mathcal{S}(\mathbb{R}^N) \to \mathcal{S}(\mathbb{R}^N)$ is linear, continuous and bijective, the operator $\mathcal{F}f = f \circ \mathcal{F}$ also admits this property. Hence the Fourier transform is an isomorphism in $\mathcal{S}^*(\mathbb{R}^N)$.

Let now $f \in \mathcal{S}^*(\mathbb{R}^N)$ and X be a Hilbert or UMD space. Then we have the norm representation

$$\|f\|_{\mathrm{H}_2^s(\mathbb{R}^N;X)} = \|(1+|\cdot|^2)^{\frac{s}{2}} \mathcal{F}f\|_{L_2(\mathbb{R}^N;X)}, \qquad s > 0. \tag{1.6}$$

If $U \subset \mathbb{R}^N$ is a subset of $\mathbb{R}^N$, then $\mathrm{H}_2^s(U;X)$ denotes the restriction of the functions $f \in \mathrm{H}_2^s(\mathbb{R}^N;X)$ to the subset U.

In case $J = [0,a]$ is an interval, we denote by ${}_0\mathrm{H}_p^s(J;X)$ the space of all functions $f : J \to X$ belonging to $\mathrm{H}_p^s(J;X)$, such that $f|_{t=0} = 0$, whenever the trace at $t = 0$ exists.

1.2. FUNCTION SPACES

By $\dot{H}_2^s(\mathbb{R}; X)$ we mean the homogenous Bessel potential space of order $s > 0$, defined as

$$\dot{H}_2^s(\mathbb{R}; X) := \left\{ f \in \mathcal{S}^*(\mathbb{R}; X) : \| |\cdot|^s \mathcal{F} f \|_{L_2(\mathbb{R};X)} < \infty \right\}. \quad (1.7)$$

By means of the fractional derivatives (1.3) and Plancherel's Theorem (cf. Theorem A.1) we obtain the identity

$$\int_\mathbb{R} |(\mathcal{F}f)(\xi)|^2 |\xi|^{2s} d\xi = \int_\mathbb{R} |\partial^s f(t)|^2 \, dt,$$

so that we have alternatively

$$\|f\|_{\dot{H}_2^s(\mathbb{R};X)} = \|\partial^s f\|_{L_2(\mathbb{R};X)}, \quad 0 < s < 2.$$

For a comprehensive account of the theory of these function spaces we refer to Triebel [106]. Observe, that (1.3), (1.6) and (1.7) allow us to define the (homogenous) Bessel potential spaces also for negative orders $s \in (-2, 0)$.

1.2.7 Weighted spaces

We will further consider weighted L_2 and W_2^s spaces. For $J := [0, a]$, $a > 0$, and a number $\mu \geq 0$ they are defined canonically via

$$L_{2,\mu}(J; X) := \{ f : J \to X : (\cdot)^\mu f \in L_2(J; X) \},$$
$$W_{2,\mu}^s(J; X) := \{ f : J \to X : (\cdot)^\mu f \in W_2^s(J; X) \}.$$

It is easy to verify that $L_2(J; X) = L_{2,0}(J; X) \hookrightarrow L_{2,\mu_1}(J; X) \hookrightarrow L_{2,\mu_2}(J; X)$ holds if and only if $\mu_1 \leq \mu_2$. With $_0W_{2,\mu}^s(J; X)$ we denote the space of all $W_{2,\mu}^s(J; X)$-functions whose trace at $t = 0$ is zero, if it exists.

Thanks to Hardy et al. [52, Theorem 329] we have the useful imbedding result

Lemma 1.1. Let V be a Banach space, $0 < \mu < 1$, and $0 < \sigma < \mu$. Then

$$_0W_{2,\mu}^\sigma(\mathbb{R}_+; V) \hookrightarrow L_{2,\mu-\sigma}(\mathbb{R}_+; V).$$

In view of homogenous Bessel potential spaces we proceed differently. We introduce the weighted homogeneous Bessel potential space $\dot{H}_2^\phi(\mathbb{R})$ with the weight function $\lambda^2 \phi(\lambda)$ as the class containing all tempered distributions $f \in \mathcal{S}^*(\mathbb{R})$ so that the number

$$\|f\|_{\dot{H}_2^\phi(\mathbb{R})} := \left[\int_\mathbb{R} |\mathcal{F}f(\lambda)|^2 \lambda^2 \phi(\lambda) \mathrm{d}\lambda \right]^{1/2} \tag{1.8}$$

is finite. It is apparent, that (1.8) defines a norm, if the function $\phi : \mathcal{D}(\phi) \to \mathbb{R}$, is almost everywhere positive and densely defined in $\mathbb{R}$. The space $\dot{H}_2^\phi(\mathbb{R})$ is an inner product space with inner product

$$(f \mid g)_{\dot{H}_2^\phi(\mathbb{R})} := \int_\mathbb{R} \mathcal{F}f(\lambda) \overline{\mathcal{F}g(\lambda)} \lambda^2 \phi(\lambda) \mathrm{d}\lambda. \tag{1.9}$$

Lemma 1.2. *If the function ϕ is even, then the inner product (1.9) of $\dot{H}_2^\phi(\mathbb{R})$ is real-valued.*

Proof. Let ϕ be even, i.e. $\phi(-\lambda) = \phi(\lambda)$ holds for every $\lambda \in \mathcal{D}(\phi)$ and recall a particular property of the Fourier transform, that is

$$\overline{\mathcal{F}f(\lambda)} = \overline{\int_\mathbb{R} e^{-it\lambda} f(t) \mathrm{d}t} = \int_\mathbb{R} e^{it\lambda} f(t) \mathrm{d}t = \mathcal{F}f(-\lambda).$$

Then, we observe

$$\begin{aligned}
(f \mid g)_{\dot{H}_2^\phi(\mathbb{R})} &= \int_\mathbb{R} \mathcal{F}f(\lambda) \overline{\mathcal{F}g(\lambda)} \lambda^2 \phi(\lambda) \mathrm{d}\lambda \\
&= \int_{-\infty}^0 \mathcal{F}f(\lambda) \overline{\mathcal{F}g(\lambda)} \lambda^2 \phi(\lambda) \mathrm{d}\lambda + \int_0^\infty \mathcal{F}f(\lambda) \overline{\mathcal{F}g(\lambda)} \lambda^2 \phi(\lambda) \mathrm{d}\lambda \\
&= \int_0^\infty \mathcal{F}f(-\lambda) \overline{\mathcal{F}g(-\lambda)} \lambda^2 \phi(\lambda) \mathrm{d}\lambda + \int_0^\infty \mathcal{F}f(\lambda) \overline{\mathcal{F}g(\lambda)} \lambda^2 \phi(\lambda) \mathrm{d}\lambda \\
&= \int_0^\infty \overline{\mathcal{F}f(\lambda)} \mathcal{F}g(\lambda) \lambda^2 \phi(\lambda) \mathrm{d}\lambda + \int_0^\infty \mathcal{F}f(\lambda) \overline{\mathcal{F}g(\lambda)} \lambda^2 \phi(\lambda) \mathrm{d}\lambda \\
&= \int_0^\infty \left[\overline{\mathcal{F}f(\lambda) \overline{\mathcal{F}g(\lambda)}} + \mathcal{F}f(\lambda) \overline{\mathcal{F}g(\lambda)} \right] \lambda^2 \phi(\lambda) \mathrm{d}\lambda \\
&= 2 \operatorname{Re} \int_0^\infty \mathcal{F}f(\lambda) \overline{\mathcal{F}g(\lambda)} \lambda^2 \phi(\lambda) \mathrm{d}\lambda
\end{aligned}$$

which yields the claim. □

1.3 Evolutionary integral equations

The notion of parabolic problems used in this study is widely taken from the monograph of Prüss [88].

Let $\mathcal{H}$ be a separable Hilbert space, A a closed linear, but in general unbounded operator in $\mathcal{H}$ with dense domain $\mathcal{D}(A)$, and let $a \in L_{1,\text{loc}}(\mathbb{R}_+)$ be of subexponential growth. Then it is readily seen that the Laplace transform $\widehat{a}(\lambda)$ of a exists for $\operatorname{Re} \lambda > 0$. We consider the Volterra equation

$$u(t) + (a * Au)(t) = f(t), \qquad t \geq 0, \tag{1.10}$$

where $f : \mathbb{R}_+ \to \mathcal{H}$ is a given function, strongly measurable and locally integrable, at least.

In the sequel we denote by $\mathcal{H}_A$ the domain of A equipped with the graph norm $|x|_A := |x| + |Ax|$. $\mathcal{H}_A$ is a Banach space since A is closed, and it is continuously embedded into $\mathcal{H}$. The following notions of solutions of (1.10) are natural. Again we let $J \subset \mathbb{R}_+$.

Definition 1.3 (Strong and mild solutions). *A function $u \in C(J; \mathcal{H})$ is called*

(a) *strong solution of (1.10) on J if $u \in C(J; \mathcal{H}_A)$ and (1.10) holds on J;*

(b) *mild solution of (1.10) on J if $a * u \in C(J; \mathcal{H}_A)$ and $u(t) = f(t) - A(a * u)(t)$ on J.*

Obviously, every strong solution of (1.10) is a mild one. The converse is not true, in general.

Definition 1.4 (Parabolicity). *Problem (1.10) is called parabolic, if*

(i) $\widehat{a}(\lambda) \neq 0$ *and* $\dfrac{1}{\widehat{a}(\lambda)} \in \rho(A)$ *for all $\operatorname{Re} \lambda > 0$;*

(ii) *there is a constant $M \geq 1$ such that*

$$\left| \frac{1}{\lambda}(I + \widehat{a}(\lambda) A)^{-1} \right| \leq \frac{M}{|\lambda|} \quad \text{for all } \operatorname{Re} \lambda > 0.$$

The notion of sectorial kernels is given by

Definition 1.5 (Sectoriality). *Let $a \in L_{1,\mathrm{loc}}(\mathbb{R}_+)$ be of subexponential growth and suppose $\hat{a}(\lambda) \neq 0$ for all $\operatorname{Re}\lambda > 0$. a is called sectorial with angle $\theta > 0$ (or merely θ-sectorial) if*

$$|\arg \hat{a}(\lambda)| \leq \theta \quad \text{for all } \operatorname{Re}\lambda > 0. \tag{1.11}$$

Here, $\arg \hat{a}(\lambda)$ is defined as the imaginary part of a fixed branch of $\log \hat{a}(\lambda)$, and θ in (1.11) is allowed to be greater than π. In case a is sectorial, we always choose that branch of $\log \hat{a}(\lambda)$ which yields the smallest angle θ; in particular, if $\hat{a}(\lambda)$ is real for real λ we choose the principal branch. In the following, we denote by $\Sigma(\omega, \theta)$ the open sector in the complex plane with vertex $\omega \in \mathbb{R}$ and opening angle 2θ which is symmetric with respect to the real positive axis. A standard situation leading to parabolic equations is described in

Proposition 1.6 ([88, Proposition 3.1]). *Let $a \in L_{1,\mathrm{loc}}(\mathbb{R}_+)$ be θ-sectorial for some $\theta < \pi$, suppose A is closed linear densely defined, such that $\rho(A) \supset \Sigma(0, \theta)$, and*

$$|(\mu + A)^{-1}| \leq \frac{M}{|\mu|} \quad \text{for all } \mu \in \Sigma(0, \theta).$$

Then (1.10) is parabolic.

The next definition introduces an appropriate notion concerning regularity of kernels.

Definition 1.7 (k-regular kernels). *Let $a \in L_{1,\mathrm{loc}}(\mathbb{R}_+)$ be of subexponential growth and $k \in \mathbb{N}$. a is called k-regular if there is a constant $c > 0$ such that*

$$|\lambda^n \hat{a}^{(n)}(\lambda)| \leq c|\hat{a}(\lambda)|, \quad \text{for all } \operatorname{Re}\lambda > 0,\ 0 \leq n \leq k.$$

It is not difficult to see that convolutions of k-regular kernels are again k-regular. Furthermore, k-regularity is preserved by integration and differentiation, while sums

1.3. EVOLUTIONARY INTEGRAL EQUATIONS

and differences of k-regular kernels need not be k-regular. However, if $a(t)$ and $b(t)$ are k-regular and

$$|\arg \widehat{a}(\lambda) - \arg \widehat{b}(\lambda)| \leq \theta < \pi, \quad \operatorname{Re}\lambda > 0$$

then $a(t)+b(t)$ is k-regular as well. In general, nonnegative, nonincreasing kernels are not 1-regular, but if the kernel is also convex, then it is 1-regular (cf. [88, Section I.3]). We call a kernel $a \in L_{1,\mathrm{loc}}(\mathbb{R}_+)$ 1-monotone if $a(t)$ is nonnegative and nonincreasing; for $k \geq 2$ we define

Definition 1.8 (k-monotone kernels). *Let $a \in L_{1,\mathrm{loc}}(\mathbb{R}_+)$ and $k \geq 2$. $a(t)$ is called k-monotone if $a \in C^{k-2}(0, \infty)$, $(-1)^n a^{(n)}(t) \geq 0$ for all $t > 0$, $0 \leq n \leq k-2$, and $(-1)^{k-2} a^{(k-2)}(t)$ is nonincreasing and convex.*

Proposition 1.9 ([88, Proposition 3.3]). *Let $k \geq 1$ and suppose $a \in L_{1,\mathrm{loc}}$ is $(k+1)$-monotone. Then $a(t)$ is k-regular and of positive type, i.e. $\frac{\pi}{2}$-sectorial.*

If A is sectorial with angle ϕ_A (for a detailed survey we refer to Denk et al. [34, Section 1]), and a is ϕ_a-sectorial, then (1.10) is parabolic provided that $\phi_A + \phi_a < \pi$, cf. [90, Proposition 3.1]. An important property of parabolic Volterra equations is the fact that they admit bounded resolvents whenever the kernel a is 1-regular, see [90, Theorem 3.1]. By a resolvent for (1.10) we mean a family $\{S(t)\}_{t \geq 0}$ of bounded linear operators in $\mathcal{H}$ which satisfy the following conditions:

(S1) $S(t)$ is strongly continuous on $\mathbb{R}_+$ and $S(0) = I$;

(S2) $S(t)\mathcal{D}(A) \subset \mathcal{D}(A)$ and $AS(t)x = S(t)Ax$ for all $x \in \mathcal{D}(A)$, $t \geq 0$;

(S3) $S(t)x + A(a * Sx)(t) = x$, for all $x \in \mathcal{H}$, $t \geq 0$.

(S3) is called resolvent equation. One can show that (1.10) admits at most one resolvent, and if it exists, then (1.10) has a unique mild solution u represented by the variation of parameters formula

$$u(t) = \frac{\mathrm{d}}{\mathrm{d}t} \int_0^t S(t-\tau) f(\tau) \mathrm{d}\tau, \quad t \geq 0, \tag{1.12}$$

at least for such f for which (1.12) is meaningful. If (1.10) admits an analytic resolvent $S(t)$ (cf. [88, Section I.1 and I.2]) which is bounded on some sector $\Sigma(0,\theta)$, then (1.10) is parabolic; the converse is not true in general.

Chapter 2

Processes with stationary increments

The theory of random processes is a very important and advanced part of modern probability theory, which is interesting from the mathematical point of view and has many applications. In practise, one has to deal particularly often with the special case of stationary random processes. Such processes naturally arise when one considers a series of observations which depend on the real-valued or integer-valued argument t (time) and do not undergo any systematic changes, but only fluctuate in a disordered manner about some constant mean level. Stationary time series can be expressed as the increment function of a process with stationary increments and occur in nearly all areas of modern technology as well as in the physical and geophysical sciences, mechanics, economics, biology and medicine, and also in many other applied fields.

This chapter is devoted to collect some fundamental definitions and regularity results, and to present an innovative approach to construct isometries of Itô-type for stochastic integrals with respect to processes X with stationary increments and spectral density. As a general reference to the topic of stationary processes we refer to Yaglom [110] and the references therein.

2.1 Definitions and Properties

Let $(\Omega, \mathcal{F}, \mathbb{P})$ be a probability space and let J be an interval of $\mathbb{R}$. An arbitrary family $\{X(t)\}_{t \in J}$ defined on Ω, such that $X(t) : \Omega \to \mathbb{R}$ is $\mathcal{F}$-measurable for each $t \in J$ is called a stochastic process and we set $X(t, \omega) = X(t)(\omega)$ for all $t \in J$ and $\omega \in \Omega$. The functions $X(\cdot, \omega)$ are called trajectories of X. For the reader's convenience we recall some basic definitions of regularity for a process $\{X(t)\}_{t \in J} \subset L_2(\Omega)$.

(a) X is mean-square continuous at $t_0 \in J$, if

$$\lim_{t \to t_0} \mathbb{E}[|X(t) - X(t_0)|^2] = 0.$$

(b) X is mean-square continuous on J, if it is mean-square continuous at every point of J.

(c) X is continuous (with probability 1), if its trajectories $X(\cdot, \omega)$ are continuous almost surely.

(d) X is Hölder-continuous of order α (with probability 1), if its trajectories $X(\cdot, \omega)$ are Hölder-continuous of order α almost surely.

Definition 2.1 (Stationary processes). *The random process X is called stationary if all its finite-dimensional distributions (or probability densities) remain the same when shifted along the time axis, that is if*

$$\{X(t)\}_{t \in \mathbb{R}} \stackrel{d}{=} \{X(t + \tau)\}_{t \in \mathbb{R}}$$

holds for all $\tau \in \mathbb{R}$. Here "$\stackrel{d}{=}$" denotes the equality in the finite-dimensional distributions.

The physical meaning of stationarity is quite clear: "It means that a phenomenon, whose numerical characteristic is the random process X, is stationary in the sense that none of the observed macroscopic factors influencing this phenomenon change

2.1. DEFINITIONS AND PROPERTIES

in time. In other words, X describes the time variation of some characteristics of a steady-state phenomenon, for which no choice of the time has any advantage over any other choice." [110, Page 52]. However, in this thesis we will not only focus on stationary motions, but rather on processes with stationary increments. In what follows we denote by

$$D_3(t; u, v) := \mathbb{E}[(X(u) - X(t))(X(v) - X(t))]$$

the structure function of the real-valued process $\{X(t)\}_{t \in J} \subset L_2(\Omega)$.

Definition 2.2 (Processes with stationary increments). *We call the random process $X := \{X(t)\}_{t \in \mathbb{R}} \subset L_2(\Omega)$ a process with stationary increments if*

(i) *the mean value of its increments depends only on the length of the incremental interval, i.e.*

$$\mathbb{E}[X(t) - X(s)] = \mathbb{E}[X(t-s) - X(0)];$$

(ii) *for $u, v, t \in \mathbb{R}$ the structure function $D_3(t; u, v)$ depends only on the differences $u - t$ and $v - t$, i.e.*

$$D_3(t; u, v) = D_3(0; u-t, v-t) =: D_2(u-t, v-t).$$

At this point the experienced reader may object that Definition 2.2 does not reflect the description of a process with stationary increments in the narrow sense, that is if

$$\{X(t) - X(s)\}_{t,s \in \mathbb{R}} \stackrel{d}{=} \{X(t+\tau) - X(s+\tau)\}_{t,s \in \mathbb{R}}$$

holds for any $\tau \in \mathbb{R}$. Therefor we should be more careful and say that the processes under consideration have stationary increments in the wider sense. However this refinement is unnecessary in this thesis where more special processes with strictly stationary increments will not be considered at all. The concept of a random process with stationary increments was introduced in Kolmogorov [65], who showed that in terms of the geometry of the Hilbert space $L_2(\Omega)$, a process with stationary increments in the manner of Definition 2.2 is in a certain sense equivalent to a screw curve.

Definition 2.2 yields that a real-valued process X with stationary increments is characterized by a function (the mean of the increments) of one variable

$$\mathbb{E}[X(t+\tau) - X(t)] =: m(\tau) \tag{2.1}$$

and by a function $D(\cdot)$ of one variable

$$\mathbb{E}|X(t+\tau) - X(t)|^2 =: D(\tau). \tag{2.2}$$

The function $D_2(\cdot, \cdot)$ can then be obtained via the identity

$$D_2(\tau_1, \tau_2) = \frac{1}{2}[D(\tau_1) + D(\tau_2) - D(|\tau_1 - \tau_2|)]. \tag{2.3}$$

Definition 2.3 (Centered processes). *A process $X := \{X(t)\}_{t \in J}$ is called centered, if $\mathbb{E}[X(t)] = 0$ holds for all $t \in J$.*

Remark 2.4. Observe, that if the process X is centered and $X(0) = 0$ a.s., then

$$D(\tau) = \operatorname{Var}[X(\tau)] \quad \text{and} \quad D_2(\tau_1, \tau_2) = \operatorname{Cov}[X(\tau_1), X(\tau_2)].$$

Looking for a general form of the function $D(\tau) = D_2(\tau, \tau)$ we follow Yaglom [110, Chapter 4] and begin with the case of differentiable processes X. This case is rather simple, since if X is a process with stationary increments and its mean square derivative $\dot{X}$ exists, this derivative clearly is a stationary process. Therefore the study of differentiable processes with stationary increments can always be reduced to the study of stationary processes $Y(t) := \dot{X}(t)$. Let

$$Y(t) = \int_{-\infty}^{\infty} e^{it\lambda} dZ_Y(\lambda), \qquad B_Y(\tau) = \int_{-\infty}^{\infty} e^{i\tau\lambda} d\Phi_Y(\lambda)$$

be spectral representations of the process Y itself and of its correlation function B_Y defined by $B_Y := \mathbb{E}[Y(t+\tau)Y(t)]$. Hereby Φ_Y is a bounded nondecreasing function, and Z_Y is a random function with uncorrelated increments, such that

$$\mathbb{E}|dZ_Y(\lambda)|^2 = d\Phi_Y(\lambda); \tag{2.4}$$

2.1. DEFINITIONS AND PROPERTIES

cf. [110, formulae (2.76) & (2.77)] in case of eventually confusions regarding the meaning. Then it is readily seen that

$$X(t) - X(0) = \int_0^t Y(\tau)\mathrm{d}\tau = \int_{-\infty}^{\infty} \left[\int_0^t e^{i\tau\lambda}\mathrm{d}\tau\right] \mathrm{d}Z_Y(\lambda).$$

Hence,

$$X(t) = \int_{-\infty}^{\infty} \frac{e^{it\lambda} - 1}{i\lambda} \mathrm{d}Z_Y(\lambda) + X(0), \tag{2.5}$$

$$D_2(\tau_1, \tau_2) = \int_{-\infty}^{\infty} \frac{(e^{i\tau_1\lambda} - 1)\overline{(e^{i\tau_2\lambda} - 1)}}{\lambda^2} \mathrm{d}\Phi_Y(\lambda),$$

and therewith

$$D(\tau) = D_2(\tau, \tau) = 2\int_{-\infty}^{\infty} \frac{1 - \cos\lambda\tau}{\lambda^2} \mathrm{d}\Phi_Y(\lambda). \tag{2.6}$$

Note that, if the point $\lambda = 0$ is a jump discontinuity of Z_Y, i.e.

$$\lim_{\varepsilon \to 0}[Z_Y(\varepsilon) - Z_Y(-\varepsilon)] = \xi \neq 0,$$

where ξ is a random variable, then due to (2.4)

$$\lim_{\varepsilon \to 0}[\Phi_Y(\varepsilon) - \Phi_Y(-\varepsilon)] = \mathbb{E}|\xi|^2 > 0,$$

i.e. the point $\lambda = 0$ is a jump discontinuity of Φ_Y also. The contribution of this discontinuity on the integral (2.6) is evidently equal to

$$2\mathbb{E}|\xi|^2 \lim_{\lambda \to 0} \frac{1 - \cos\lambda\tau}{\lambda^2} = \mathbb{E}|\xi|^2 \tau^2.$$

Let us further introduce the functions Z and Φ by

$$Z(\lambda_2) - Z(\lambda_1) = \int_{\lambda_1}^{\lambda_2} \frac{\mathrm{d}Z_Y(\lambda)}{i\lambda}, \qquad \Phi(\lambda_2) - \Phi(\lambda_1) = \int_{\lambda_1}^{\lambda_2} \frac{\mathrm{d}\Phi_Y(\lambda)}{\lambda^2} \tag{2.7}$$

for $0 < \lambda_1 < \lambda_2$ or $\lambda_1 < \lambda_2 < 0$. Then, considering again the real-valued case and interpreting the integral over $\mathbb{R}$ as the limit

$$\int_{-\infty}^{\infty} = \lim_{R \to \infty, \varepsilon \to 0} \left\{\int_{-R}^{-\varepsilon} + \int_{\varepsilon}^{R}\right\},$$

formulae (2.5) and (2.6) can also be written as

$$X(t) = \int_{-\infty}^{\infty} (e^{it\lambda} - 1)\mathrm{d}Z(\lambda) + X(0) + \xi t, \tag{2.8}$$

respectively

$$D(\tau) = 4\int_0^\infty (1-\cos\lambda\tau)\mathrm{d}\Phi(\lambda) + \mathbb{E}|\xi|^2\tau^2. \tag{2.9}$$

By virtue of (2.7) the spectral distribution Φ is a nondecreasing function on the half-lines $(-\infty,0)$ and $(0,\infty)$ such that

$$\int_{-\infty}^\infty \lambda^2\mathrm{d}\Phi(\lambda) < \infty. \tag{2.10}$$

If the correlation function $B_Y(\tau)$ of $Y = \dot{X}$ falls off rapidly enough with $|\tau|$ (say $B_Y \in L_2(\mathbb{R})$), then $\mathbb{E}|\xi|^2 = 0$ and $\mathrm{d}\Phi(\lambda)$ can be replaced in (2.9) by $\phi(\lambda)\mathrm{d}\lambda$, where $\phi(\lambda) = \Phi'(\lambda) \geq 0$ is the spectral density of the process X. Thus, if the spectral density exists (say when Φ is absolutely continuous), we can rewrite formula (2.9) into

$$D(\tau) = 4\int_0^\infty (1-\cos\lambda\tau)\phi(\lambda)\mathrm{d}\lambda, \tag{2.11}$$

where

$$\int_0^\infty \lambda^2\phi(\lambda)\mathrm{d}\lambda < \infty.$$

It can be shown (e.g. von Neumann & Schoenberg [109]) that spectral representations similar to the above also exists for any nondifferentiable process X with stationary increments, with the only difference that, in general, the spectral distribution $\Phi(\lambda)$ increases so rapidly as $|\lambda| \to \infty$, that the integral (2.10) becomes infinite. Then instead of (2.10) it is only necessary that, for any $\lambda_0 > 0$,

$$\int_{-\infty}^{-\lambda_0} \mathrm{d}\Phi(\lambda) + \int_{-\lambda_0}^{\lambda_0} \lambda^2\mathrm{d}\Phi(\lambda) + \int_{\lambda_0}^\infty \mathrm{d}\Phi(\lambda) < \infty.$$

If, as in this thesis, the processes X is real and its spectral density exists, then $\phi(\lambda) = \phi(-\lambda)$ and, for any $\lambda_0 > 0$ it is

$$\int_0^{\lambda_0} \lambda^2\phi(\lambda)\mathrm{d}\lambda + \int_{\lambda_0}^\infty \phi(\lambda)\mathrm{d}\lambda < \infty, \tag{2.12}$$

but the integral $\int_0^\infty \phi(\lambda)\mathrm{d}\lambda$ may be infinite, if X has stationary increments but is not stationary itself. We have shown that if X is a process with stationary increments, then X and its structure function $D(\cdot)$ have the spectral representations (2.8) and (2.9), respectively. Von Neumann & Schoenberg [109] proved that the converse is also true.

Before turning to regularity, we will give a precise meaning to the spectrum of the process X. That is the frequency λ is said to belong to the spectrum of X if $\Phi(\lambda + \varepsilon) - \Phi(\lambda - \varepsilon) > 0$ for any $\varepsilon > 0$, where Φ denotes the spectral distribution of X. If X possesses a spectral density ϕ, then the spectrum of X is the closure of the set $\{\lambda \in \mathbb{R} : \phi(\lambda) > 0\}$, or in other words, the spectrum consists of all frequencies λ which have no vicinity where the spectral density ϕ identically vanishes.

We will, in the subsequent section, characterize two classes of processes with stationary increments. As a first goal we then accomplish to derive estimates on the structure function D in order to deduce results concerning regularity in the pathwise and in the $L_p(\Omega)$-sense.

2.2 Regularity

From now on we will exclude the case $\phi \equiv 0$, since this case merely corresponds to the trivial process $X \equiv 0$ a.s. In view on regularity results, we may classify the processes under consideration with respect to their spectral densities. The first class contains all processes X which satisfy

Hypothesis (ϕ). X is a real-valued process with stationary increments and $X(0) = 0$ a.s. Furthermore, the spectral density ϕ of X exists and there is a number $1 < \gamma < 3$, so that

$$\sup_{\lambda \in \mathbb{R}} |\lambda|^\gamma \phi(\lambda) < \infty.$$

Observe that Hypothesis (ϕ) does not directly incorporate the spectrum of X. The second class is abstractly formulated as

Hypothesis (ϕ_0). X is a real-valued process with stationary increments and $X(0) = 0$ a.s. Furthermore, the spectral density ϕ of X exists and there are numbers $1 < \gamma_0 < 3$, $\lambda_0 > 0$ and $\theta \geq 0$, so that

(a) $\inf_{0 < \lambda < \lambda_0} \lambda^{\gamma_0} \phi(\lambda) > 0$,

(b) $\limsup_{\lambda \to 0} |\lambda|^{\gamma_0} \phi(\lambda) < \infty$,

(c) $\phi(\tau \lambda) \geq \tau^{-(\gamma_0 + \theta)} \phi(\lambda)$ for all $\lambda \in (0, \infty)$ and $\tau \geq 1$.

This particulary means the frequency zero must necessarily be contained in the spectrum of the process X.

Remark 2.5. Note the following.

1. The restrictions $\gamma_0 < 3$ and $\theta \geq 0$ are evident since the spectral density ϕ must satisfy condition (2.12). Moreover, the restriction $\gamma < 3$ (resp. $\gamma > 1$) is nontrivial if $\lambda = 0$ is not contained in the spectrum of X (resp. the spectrum of X is bounded).

2. In view of applications one should always be exerted to choose the number θ preferably small in order to achieve optimal regularity results (see Theorems 2.18 and 2.21 below).

3. Suppose the process X is subject to Hypothesis (ϕ_0) with the number θ chosen to be as small as possible.

    (a) For any fixed $h > 0$, the spectral density $\phi_h(\lambda) = 2(1 - \cos \lambda h) \phi(\lambda)$ of the incremental process $X_h(t) := [X(t+h) - X(t)]$ has a singularity at frequency zero if and only if $\gamma_0 > 2$. It is frequently claimed in literature, that in this case X displays long-range dependence in the sense that the dependence between the increments $[X(1) - X(0)]$ and $[X(n+1) - X(n)]$ decays slowly as n tends to infinity and
    $$\sum_{n=1}^{\infty} \mathrm{Cov}[X(1) - X(0), X(n+1) - X(n)] = \infty.$$

2.2. REGULARITY

However, we stress that this is not true in general (cf. Gubner [51]).

(b) If $\theta > 0$, then ϕ has a significance in its behavior when $|\lambda| \to \infty$. As a consequence the correlation of consecutive small increments of X exceeds the correlation of consecutive large increments (cf. Remark 2.17 below). This phenomena is called intermittency in turbulence literature (e.g. Frisch [44]).

Example 2.6. Suppose X is a process with stationary increments and $X(0) = 0$ a.s. Assume the spectral density ϕ exists and is of the form

$$\phi(\lambda) = \frac{1}{|\lambda|^\alpha (1+|\lambda|^s)^\beta}, \quad 1 < \alpha < 3, \quad s \geq 0, \quad \beta \geq 0.$$

Then X satisfies Hypothesis (ϕ), whereby the number γ can be chosen in $[\alpha, \alpha + s\beta] \cap [\alpha, 3)$, since for this selection

$$|\lambda|^\gamma \phi(\lambda) = \frac{1}{|\lambda|^{\alpha-\gamma}(1+|\lambda|^s)^\beta}$$

is clearly bounded on $\mathbb{R}$. Moreover, X is due to Hypothesis (ϕ_0) with $\gamma_0 = \alpha$ and $\theta \geq s\beta$. This is apparent because

$$|\lambda|^{\gamma_0} \phi(\lambda) = \frac{1}{|\lambda|^{\alpha-\gamma_0}(1+|\lambda|^s)^\beta}$$

is strictly positive and bounded in a neighborhood of $\lambda = 0$ if and only if $\gamma_0 = \alpha$ and

$$\phi(\tau\lambda) = \frac{1}{(\tau\lambda)^{\gamma_0}(1+|\tau\lambda|^s)^\beta} = |\tau|^{-(\gamma_0+s\beta)} \frac{1}{\lambda^{\gamma_0}(|\tau|^{-s}+|\lambda|^s)^\beta}$$

$$\geq |\tau|^{-(\gamma_0+s\beta)} \frac{1}{\lambda^{\gamma_0}(1+|\lambda|^s)^\beta} = |\tau|^{-(\gamma_0+s\beta)} \phi(\lambda)$$

obviously holds true for all $|\tau| \geq 1$.

As we have seen in the previous example, there are processes which are due to both, Hypotheses (ϕ) and (ϕ_0). However, this is not true in general.

Example 2.7. Consider a process X with stationary increments being zero at time zero a.s. and suppose ϕ is of the form

$$\phi(\lambda) = \frac{1 - \sin(|\lambda|)}{|\lambda|^\alpha}, \quad 1 < \alpha < 3.$$

Then X is subject to Hypothesis (ϕ) with $\gamma = \alpha$, but $|\lambda|^\gamma \phi(\lambda) = 1 - \sin(|\lambda|)$ violates the growth condition (c) of Hypothesis (ϕ_0) for any $\theta \geq 0$.

Vice versa, we find processes which satisfy Hypothesis (ϕ_0), but conflict (ϕ).

Example 2.8. Consider a spectral density of the form

$$\phi(\lambda) = \frac{1 + |\lambda|^s}{|\lambda|^\alpha}, \quad 2 < \alpha < 3, \quad 0 < s < \alpha - 1.$$

Then the associated process X might satisfy Hypothesis (ϕ_0) with $\gamma_0 = \alpha$ and $\theta = 0$, but will surely contradict Hypothesis (ϕ). This is due to the fact, that the singularity at frequency zero compensates a singularity at infinity and can therefore not factorized as a power of $|\lambda|$, so that the remainder becomes bounded.

The following proposition clarifies, how Hypotheses (ϕ) and (ϕ_0) are connected. For brevity we define

$$f(\lambda) := |\lambda|^\gamma \phi(\lambda), \quad g(\lambda) := |\lambda|^{\gamma_0} \phi(\lambda), \quad S_f := \sup_{\lambda \in \mathbb{R}} f(\lambda), \quad I_g := \inf_{0 < \lambda < \lambda_0} g(\lambda). \quad (2.13)$$

Note that f, g, S_f and I_g depend on the parameters γ, γ_0 and λ_0, respectively.

Proposition 2.9. *Let X satisfy Hypotheses (ϕ) and (ϕ_0), then $\gamma_0 \leq \gamma \leq \gamma_0 + \theta$ and $S_f \geq \lambda_0^{\gamma - \gamma_0} I_g$. Moreover, $\theta = 0$ is equivalent to $f = g$. If this is the case, then the remainders are nondecreasing on the half-line $(0, \infty)$ and satisfy*

$$I_g = \lim_{\lambda \to 0} f(\lambda) \leq \lim_{|\lambda| \to \infty} f(\lambda) = S_f.$$

Proof. Observe the estimate for $|\tau| \geq 1$ and $\lambda \in (0, \lambda_0)$

$$|\lambda|^{-\gamma} |\tau|^{-\gamma} S_f \geq |\lambda|^{-\gamma} |\tau|^{-\gamma} f(\tau \lambda) \geq |\lambda|^{-\gamma_0} |\tau|^{-\gamma_0 - \theta} g(\lambda) \geq |\lambda|^{-\gamma_0} |\tau|^{-\gamma_0 - \theta} I_g,$$

2.2. REGULARITY

in particular
$$\frac{S_f}{I_g} \geq |\lambda|^{\gamma-\gamma_0}|\tau|^{\gamma-\gamma_0-\theta}.$$
Thus necessarily $\gamma_0 \leq \gamma \leq \gamma_0 + \theta$ and $S_f \geq \lambda_0^{\gamma-\gamma_0} I_g$. The case $\theta = 0$ corresponds to $\gamma_0 = \gamma$ and therewith $f = g$. If this is the case then f is obviously bounded and $f(\lambda) \geq I_g$ in a neighborhood of $\lambda = 0$. Moreover f satisfies the growth condition $f(\tau\lambda) \geq f(\lambda)$ for all $\lambda > 0$ and $\tau \geq 1$, thus f is nondecreasing on $(0, \infty)$ and, since f is an even function, nonincreasing on $(-\infty, 0)$. □

Corollary 2.10. *Let X satisfy Hypotheses (ϕ) and (ϕ_0) with $\theta > 0$. Then the choice $\gamma = \gamma_0$ is admissible. If, in addition, $\limsup_{\lambda \to \infty} \lambda^\theta g(\lambda) < \infty$, then any selection $\gamma \in [\gamma_0, \gamma_0 + \theta] \cap [\gamma_0, 3)$ is feasible.*

Proof. Recall, that by Proposition 2.9 we have $\gamma_0 \leq \gamma \leq \gamma_0 + \theta$. To justify the choice $\gamma = \gamma_0$ we have to show that $\sup_{\lambda > 0} \lambda^{\gamma_0} \phi(\lambda) < \infty$. This can be seen from
$$|\lambda|^{\gamma_0-\gamma} f(\lambda) = |\lambda|^{\gamma_0} \phi(\lambda) = g(\lambda).$$
Turning to the second claim, we need to verify $\sup_{\lambda > 0} \lambda^{\gamma_0+\theta} \phi(\lambda) < \infty$, which follows from
$$|\lambda|^{\gamma_0+\theta-\gamma} f(\lambda) = |\lambda|^{\gamma_0+\theta} \phi(\lambda) = |\lambda|^\theta g(\lambda).$$
□

Theorem 2.11. *The following are true.*

(i) Let X be subject to Hypothesis (ϕ). Then the estimate
$$D(\tau) \leq c_\phi |\tau|^{\gamma-1}, \qquad c_\phi := c_\phi(\gamma) = 2^{4-\gamma} \int_0^\infty \frac{\sin^2(\lambda)}{\lambda^\gamma} d\lambda \cdot \sup_{\lambda \in \mathbb{R}} |\lambda|^\gamma \phi(\lambda) \qquad (2.14)$$
holds for all $\tau \in \mathbb{R}$. Moreover, (2.14) holds with equality if $|\lambda|^\gamma \phi(\lambda)$ is identically constant.

(ii) Let X be subject to Hypothesis (ϕ_0). Then the estimate
$$D(\tau) \geq c_{\phi_0} \cdot \min\{|\tau|^{\gamma_0-1+\theta}, |\tau|^{\gamma_0-1}\},$$
$$c_{\phi_0} := c_{\phi_0}(\gamma_0, \lambda_0) = 2^{4-\gamma_0} \int_0^{\lambda_0/2} \frac{\sin^2(\lambda)}{\lambda^{\gamma_0}} d\lambda \cdot \inf_{|\lambda| < \lambda_0} |\lambda|^{\gamma_0} \phi(\lambda)$$

holds for all $\tau \in \mathbb{R}$.

Proof. It is particularly seen from (2.11) that $D(\tau) = D(-\tau)$ which entails, that this proof can be reduced to the case $\tau \geq 0$. If $\tau = 0$ then trivially $D(\tau) = 0$ so that it suffices to prove the claim for $\tau > 0$. The results then follow from the spectral representation (2.11) of the function D, because for all $\tau > 0$ it is

$$D(\tau) = 4\int_0^\infty (1-\cos\xi\tau)\phi(\xi)\mathrm{d}\xi = 8\int_0^\infty \sin^2\left(\frac{\xi\tau}{2}\right)\phi(\xi)\mathrm{d}\xi = \frac{16}{\tau}\int_0^\infty \sin^2(\lambda)\phi\left(\frac{2\lambda}{\tau}\right)\mathrm{d}\lambda.$$

Observe now that $\int_0^\infty \sin^2(\lambda)/\lambda^\alpha \mathrm{d}\lambda$ exists if $1 < \alpha < 3$. Then by means of notations (2.13), assertion (i) follows with

$$D(\tau) = 2^{4-\gamma}\tau^{\gamma-1}\int_0^\infty \frac{\sin^2\lambda}{\lambda^\gamma}f\left(\frac{2\lambda}{\tau}\right)\mathrm{d}\lambda \leq \left[2^{4-\gamma}S_f\int_0^\infty \frac{\sin^2\lambda}{\lambda^\gamma}\mathrm{d}\lambda\right]\tau^{\gamma-1} = c_\phi\tau^{\gamma-1},$$

while for $0 < \tau < 1$ (ii) is a consequence of the growth condition (ϕ_0)(c), because

$$D(\tau) = 2^{4-\gamma_0}\tau^{\gamma_0-1}\int_0^\infty \frac{\sin^2\lambda}{\lambda^{\gamma_0}}g\left(\frac{2\lambda}{\tau}\right)\mathrm{d}\lambda \geq \left[2^{4-\gamma_0}I_g\int_0^{\lambda_0/2}\frac{\sin^2\lambda}{\lambda^{\gamma_0}}\mathrm{d}\lambda\right]\tau^{\gamma_0-1+\theta} = c_{\phi_0}\tau^{\gamma_0-1+\theta}.$$

In case $\tau \geq 1$ assertion (ii) follows with strict positivity of the remainder g in a neighborhood of $\lambda = 0$. Then

$$D(\tau) = 2^{4-\gamma_0}\tau^{\gamma_0-1}\int_0^\infty \frac{\sin^2\lambda}{\lambda^{\gamma_0}}g\left(\frac{2\lambda}{\tau}\right)\mathrm{d}\lambda \geq \left[2^{4-\gamma_0}I_g\int_0^{\lambda_0/2}\frac{\sin^2\lambda}{\lambda^{\gamma_0}}\mathrm{d}\lambda\right]\tau^{\gamma_0-1} = c_{\phi_0}\tau^{\gamma_0-1}.$$

□

Following the latter proof we may outline, that the growth condition (c) of Hypothesis (ϕ_0) is only involved when $|\tau| < 1$ and that the constant c_ϕ particularly depends on the parameter γ. This is crucial to remember in situations where the parameter γ is not uniquely determined by Hypothesis (ϕ). As an immediate consequence of Theorem 2.11(i) we obtain

Corollary 2.12. *Suppose X be subject to Hypothesis (ϕ) and denote by Γ the set of all feasible γ. Then the estimate*

$$D(\tau) \leq \inf\left\{c_\phi(\gamma)|\tau|^{\gamma-1} : \gamma \in \Gamma\right\}$$

holds for all $\tau \in \mathbb{R}$, with $c_\phi := c_\phi(\gamma)$ from Theorem 2.11(i).

2.2. REGULARITY

The result of Theorem 2.11(ii) is very convenient with view on applications. The downside of the simple representation is clearly the fact, that it is not optimal in general. The following corollary can be deduced in similar fashion as of Theorem 2.11(ii) and yields a lower bound for the second moment of $X(t)$ which performs much better.

Corollary 2.13. Let X be subject to Hypothesis (ϕ_0) and denote by Λ the set of all admissible λ_0. Then the estimate

$$D(\tau) \geq \sup_{R>0, \lambda_0 \in \Lambda} \left[c_R \left(\inf_{|\lambda| < \lambda_0} g(\lambda) \right) \cdot \begin{cases} \left(\frac{\lambda_0}{2R}\right)^\theta \tau^{\gamma_0 + \theta - 1} & : \tau \leq \frac{2R}{\lambda_0} \\ \tau^{\gamma_0 - 1} & : \tau > \frac{2R}{\lambda_0} \end{cases} \right], \quad c_R := 2^{4-\gamma_0} \int_0^R \frac{\sin^2 \lambda}{\lambda^{\gamma_0}} d\lambda$$

holds true for all $\tau \in \mathbb{R}$.

Proof. Follow the lines of the proof of Theorem 2.11(ii) to verify

$$D(\tau) = 2^{4-\gamma_0} \tau^{\gamma_0 - 1} \int_0^\infty \frac{\sin^2 \lambda}{\lambda^{\gamma_0}} g\left(\frac{2\lambda}{\tau}\right) d\lambda.$$

By now we fix an arbitrary $\lambda_0 \in \Lambda$ and a number $R > 0$ to proceed with

$$D(\tau) \geq 2^{4-\gamma_0} \tau^{\gamma_0 - 1} \int_0^R \frac{\sin^2 \lambda}{\lambda^{\gamma_0}} g\left(\frac{2\lambda}{\tau}\right) d\lambda.$$

Observe now, that $\frac{2R}{\tau} < \lambda_0$ if $\tau > \frac{2R}{\lambda_0}$ and, moreover by (ϕ_0)(c),

$$g\left(\frac{2\lambda}{\tau}\right) = g\left(\frac{2R}{\lambda_0 \tau} \cdot \frac{\lambda \lambda_0}{R}\right) \geq \left(\frac{2R}{\lambda_0 \tau}\right)^{-\theta} g\left(\frac{\lambda \lambda_0}{R}\right),$$

provided that $\tau \leq \frac{2R}{\lambda_0}$. The claim can then be enforced by employing the remaining arguments of the proof of Theorem 2.11(ii) for each pair $(R, \lambda_0) \in (0, \infty) \times \Lambda$. $\square$

If the process X is in particular centered and Gaussian, then we obtain L_p-estimates with the aid of the Kahane-Khinchine inequality (cf. Theorem A.3).

Corollary 2.14. Let X be a centered Gaussian process.

(i) If X is subject to Hypothesis (ϕ), then for every $p \in (2, \infty)$ there is a constant $c > 0$ such that the estimate
$$\mathbb{E}|X(t) - X(s)|^p \le c|t-s|^{\frac{p(\gamma-1)}{2}}$$
holds for all $t, s \in \mathbb{R}$.

(ii) If X is subject to Hypothesis (ϕ_0), then for every $p \in (1, 2)$ there is a constant $c_0 > 0$ such that the estimate
$$\mathbb{E}|X(t) - X(s)|^p \ge c_0 \cdot \min\{|t-s|^{\frac{p(\gamma_0-1+\theta)}{2}}, |t-s|^{\frac{p(\gamma_0-1)}{2}}\}$$
holds for all $t, s \in \mathbb{R}$.

Proof. With the aid of the Kahane-Khinchine inequality (cf. Theorem A.3), the claim (i) follows directly from Theorem 2.11(i) because
$$\mathbb{E}|X(t) - X(s)|^p \le c \left(\mathbb{E}|X(t) - X(s)|^2\right)^{p/2} = c\left(D(t-s)\right)^{p/2} \le c|t-s|^{p\frac{\gamma-1}{2}},$$
holds for all $p > 2$. Here the constant $c > 0$ is generic and may depend on p. The second assertion follows in a similar manner. □

We now may take a closer look to the correlation of the increments in case X is centered. We start with the case of small consecutive increments.

Proposition 2.15. Assume X is centered and satisfies Hypotheses (ϕ) and (ϕ_0) with $\gamma = \gamma_0 + \theta$. Denote by c_ϕ and c_{ϕ_0} be the constants from Theorem 2.11 and let $0 < \tau \le \frac{1}{2}$.

(i) If $\gamma_0 < 2 - \log_2(c_\phi/c_{\phi_0}) - \theta$, then the increments $[X(t) - X(t-\tau)]$ and $[X(t+\tau) - X(t)]$ are negative correlated.

(ii) If $\gamma_0 > 2 + \log_2(c_\phi/c_{\phi_0}) - \theta$, then the increments $[X(t) - X(t-\tau)]$ and $[X(t+\tau) - X(t)]$ are positive correlated.

2.2. REGULARITY

Proof. With the aid of identity (2.3), a direct computation verifies

$$\mathrm{Cov}[X(t) - X(t-\tau), X(t+\tau) - X(t)] = \mathbb{E}[(X(t) - X(t-\tau))(X(t+\tau) - X(t))]$$
$$= -\mathbb{E}[X(-\tau)X(\tau)] = \frac{1}{2}\left(\mathbb{E}[X(2\tau)] - \mathbb{E}[X(-\tau)]^2 - \mathbb{E}[X(\tau)]^2\right)$$

and it is a matter of the stationarity of the increments and the assumption $\mathbb{E}[X(0)] = 0$, that

$$\mathbb{E}[X(-\tau)]^2 = \mathbb{E}[X(0) - X(-\tau)]^2 = \mathbb{E}[X(\tau) - X(0)]^2 = \mathbb{E}[X(\tau)]^2.$$

Thus, we already have

$$\mathbb{E}[(X(t) - X(t-\tau))(X(t+\tau) - X(t))] = \frac{1}{2}\mathbb{E}[X(2\tau)]^2 - \mathbb{E}[X(\tau)]^2$$

and we may employ Theorem 2.11 to estimate

$$\frac{c_{\phi_0}}{2}(2\tau)^{\gamma_0 - 1 + \theta} - c_\phi \tau^{\gamma - 1} \leq \mathbb{E}[(X(t) - X(t-\tau))(X(t+\tau) - X(t))] \leq \frac{c_\phi}{2}(2\tau)^{\gamma - 1} - c_{\phi_0}\tau^{\gamma_0 - 1 + \theta}.$$

Recall that $\gamma = \gamma_0 + \theta$ by assumption. Thus

$$(c_{\phi_0} 2^{\gamma_0 + \theta - 2} - c_\phi)\tau^{\gamma_0 + \theta - 1} \leq \mathbb{E}[(X(t) - X(t-\tau))(X(t+\tau) - X(t))] \leq (c_\phi 2^{\gamma_0 + \theta - 2} - c_{\phi_0})\tau^{\gamma_0 + \theta - 1}.$$

and the result is immediate. □

Following the same strategy one observes a result for large increments of X.

Proposition 2.16. *Assume X is centered and satisfies Hypotheses (ϕ) and (ϕ_0). Denote by c_ϕ and c_{ϕ_0} the constants from Theorem 2.11 and let $\tau \geq 1$.*

(i) *If $\gamma_0 < 2 - \log_2(c_\phi/c_{\phi_0})$, then the increments $[X(t) - X(t-\tau)]$ and $[X(t+\tau) - X(t)]$ are negative correlated.*

(ii) *If $\gamma_0 > 2 + \log_2(c_\phi/c_{\phi_0})$, then the increments $[X(t) - X(t-\tau)]$ and $[X(t+\tau) - X(t)]$ are positive correlated.*

Remark 2.17. Propositions 2.15 and 2.16 show in particular, that if X is centered and satisfies Hypotheses (ϕ) and (ϕ_0) with $\theta > 0$ and admissible $\gamma \in [\gamma_0, \gamma_0 + \theta]$, the correlations of small and large increments satisfy

$$\mathbb{E}[(X(t) - X(t - \tau_1))(X(t + \tau_1) - X(t))] > \mathbb{E}[(X(t) - X(t - \tau_2))(X(t + \tau_2) - X(t))],$$

with $0 < \tau_1 \leq \frac{1}{2}$ and $\tau_2 \geq 1$. A very special situation is described if $\gamma_0 < 2 - \log_2(c_\phi/c_{\phi_0})$ and $\theta > 2\log_2(c_\phi/c_{\phi_0})$. In this case large increments are negative correlated, while consecutive small increments tend to have the same sign.

The subsequent theorem yields first regularity results in the pathwise sense. Due to the available L_p-estimates for centered Gaussian processes the results can be enhanced in this case.

Theorem 2.18. *The following are true.*

(i) *Let X be subject to Hypothesis (ϕ). If $\gamma > 2$, then X is mean-square continuous and has continuous paths almost sure. Moreover, the trajectories of X are locally Hölder-continuous of order $\alpha < \frac{\gamma-2}{2}$ with probability 1.*

(ii) *Let X be a centered Gaussian process subject to Hypothesis (ϕ). Then X is mean-square continuous and has continuous paths almost sure. Moreover, the trajectories of X are locally Hölder-continuous of order $\alpha < \frac{\gamma-1}{2}$ with probability 1.*

(iii) *Let X be subject to Hypothesis (ϕ_0). If $\theta < 3 - \gamma_0$ then X is almost surely nowhere mean-square differentiable.*

Proof. Assertion (i) is immediate by Theorem 2.11(i), which in particular yields

$$\mathbb{E}[X(t) - X(s)]^2 \leq c|t-s|^{1+(\gamma-2)}, \quad t, s \in \mathbb{R}.$$

Thus the Kolmogorov-Čentsov-Theorem (cf. Theorem A.4) proves the claim. Regarding (ii) Corollary 2.14 yields, that for every $p \in (2, \infty)$ there is a constant $c > 0$ such that

$$\mathbb{E}|X(t) - X(s)|^p \leq c|t-s|^{1+\frac{p(\gamma-1)-2}{2}},$$

2.2. REGULARITY

for all $s, t \in \mathbb{R}$. Employing the Kolmogorov-Čentsov-Theorem (cf. Theorem A.4) yields the Hölder-continuity on every bounded subset of $\mathbb{R}$ of order $\alpha < \frac{\gamma-1}{2} - \frac{1}{p}$ for every $2 < p < \infty$. In case (iii) the nowhere differentiability in the $L_2(\Omega)$-sense follows with

$$\lim_{s \to t} \left| \frac{\mathbb{E}[X(t) - X(s)]^2}{|t-s|^2} \right| = \lim_{s \to t} \frac{\mathbb{E}[X(t-s)]^2}{|t-s|^2} \geq c_0 \lim_{s \to t} |t-s|^{\gamma_0 - 3 + \theta} = \infty.$$ □

Corollary 2.19. Let X be subject to Hypothesis (ϕ). If $\gamma > 2$, then X is centered.

Proof. Observe that the function m given by (2.1) satisfies

$$m(\tau_1) + m(\tau_2) = m(\tau_1 + \tau_2)$$

and, since $\gamma > 2$, the function m is continuous by Theorem 2.18(i), which in turn yields $m(\tau) = c_1 \tau$ where c_1 is a constant. Because

$$D(\tau) = (\mathbb{E}[X(t+\tau) - X(t)])^2 + \mathrm{Var}[X(t+\tau) - X(t)] = c_1^2 \tau^2 + \mathrm{Var}[X(t+\tau) - X(t)],$$

where, as a matter of course $\mathrm{Var}[X(t+\tau) - X(t)] \geq 0$, the function $D(\tau)$ includes a term proportional to τ^2 in all the cases where $c_1 \neq 0$. Theorem 2.11 yields $D(\tau) \leq c|\tau|^{\gamma-1}$ for all $\tau \in \mathbb{R}$ with $\gamma - 1 < 2$. Hence $c_1 = 0$, which is equivalent to $0 = m(\tau) = \mathbb{E}[X(t+\tau) - X(t)] = \mathbb{E}[X(\tau)]$ for all $\tau \in \mathbb{R}$. □

Corollary 2.20. Let X be a Gaussian process subject to Hypothesis (ϕ). If $\gamma > 2$, then with probability 1 the trajectories of X are locally Hölder-continuous of order $\alpha < \frac{\gamma-1}{2}$.

Proof. The result is immediate by Corollary 2.19 and Theorem 2.18(ii). □

We are now in the position to formulate first results in the $L_p(\Omega)$-sense.

Theorem 2.21. Let $T > 0$, $J = [0,T]$, $p \in (0, \infty)$ and $0 < \sigma < 1$.

(i) Suppose X satisfies Hypothesis (ϕ). If $2\sigma < \gamma - 1$, then $X \in {}_0W_2^\sigma(J; L_2(\Omega))$.

(ii) Suppose X satisfies Hypothesis (ϕ_0). If $2\sigma \geq \gamma_0 - 1 + \theta$, then $X \notin {}_0W_2^\sigma(J; L_2(\Omega))$.

(iii) Suppose X is a centered Gaussian process subject to Hypothesis (ϕ) and let $2 \leq q < \infty$. If $2\sigma < \gamma - 1$, then $X \in {}_0W_p^\sigma(J; L_q(\Omega))$.

(iv) Suppose X is a centered Gaussian process subject to Hypothesis (ϕ_0) and let $1 < q \leq 2$. If $2\sigma \geq \gamma_0 - 1 + \theta$, then $X \notin {}_0W_p^\sigma(J; L_q(\Omega))$.

Proof. In view of Theorem 2.11 and Corollary 2.14, assertions (iii) and (iv) will be shown directly considering the semi-norm $[\cdot]_\sigma$ in ${}_0W_p^\sigma(J; L_q(\Omega))$ and exploiting the stationarity of the increments. Assertion (iii) follows with

$$[X]_\sigma^p = \int_J \int_J \frac{(\mathbb{E}|X(t) - X(s)|^q)^{\frac{p}{q}}}{|t-s|^{1+p\sigma}} \mathrm{d}s\mathrm{d}t \leq c \int_J \int_J \frac{|t-s|^{\frac{p(\gamma-1)}{2}}}{|t-s|^{1+p\sigma}} \mathrm{d}s\mathrm{d}t$$

$$= c \int_J \int_J |t-s|^{\frac{p(\gamma-1)}{2} - 1 - p\sigma} \mathrm{d}s\mathrm{d}t = 2c \int_0^T \int_0^t (t-s)^{\frac{p(\gamma-1)}{2} - 1 - p\sigma} \mathrm{d}s\mathrm{d}t$$

and the last integral is finite, if and only if $\sigma < \frac{\gamma-1}{2}$. Turning to (iv) we restrict J to $[0,1]$ and achieve analogously

$$[X]_\sigma^p \geq 2c_0 \int_0^1 \int_0^t (t-s)^{\frac{p(\gamma_0-1+\theta)}{2} - 1 - p\sigma} \mathrm{d}s\mathrm{d}t$$

and the right hand-side is finite if and only if $\sigma < \frac{\gamma_0-1+\theta}{2}$. To prove assertions (i) and (ii) set $p = q = 2$ and repeat the above arguments. $\square$

2.3 Noise

In the analytical treatment it is inconvenient that a process X does in general not have a derivative. A very elementary method of circumventing the lack of derivative is to smooth X and to introduce the random function

$$X_\delta(t) := \frac{1}{\delta} \int_t^{t+\delta} X(\tau) \mathrm{d}\tau = \int_{\mathbb{R}} X(\tau) \varphi_1(t-\tau) \mathrm{d}\tau, \qquad \delta > 0,$$

where
$$\varphi_1(t) := \begin{cases} \delta^{-1} & : 0 \leq t \leq \delta, \\ 0 & : \text{otherwise.} \end{cases}$$

The process X_δ has a stationary derivative

$$\dot{X}_\delta(t) = \frac{1}{\delta}[X(t+\delta) - X(t)] = -\int_\mathbb{R} X(\tau) \mathrm{d}\varphi_1(t-\tau),$$

which is almost surely continuous, but surely nondifferentiable. For δ small enough, the processes X and X_δ are indistinguishable for many practical purposes. One can replace φ_1 by an infinitely differentiable kernel φ, which vanishes outside some finite interval and integrates to 1. Then

$$\partial_t^k \int_\mathbb{R} X(\tau)\varphi(t-\tau)\mathrm{d}\tau = (-1)^k \int_\mathbb{R} X(\tau)\varphi^{(k)}(t-\tau)\mathrm{d}\tau,$$

which is continuous and stationary for all $k \in \mathbb{N}$. Following up this approach, one can interpret

$$\dot{X}(t) = -\int_\mathbb{R} X(\tau)\dot{\varphi}(t-\tau)\mathrm{d}\tau$$

as being not a random function but a generalized random function in the sense of Schwartz distributions (e.g. Gelfand & Vilenkin [45]). For practical purposes, it may be desirable to avoid Schwartz distributions, and one shall be concerned with determining whether finite differences of the process $\dot{X}_\delta$ are reasonable approximations of the noise $\dot{X}$.

2.4 Deterministic multipliers

Throughout this section let $J = [0, T]$, $G \subset \mathbb{R}^N$, $W = L_2(J; L_2(G; L_2(\Omega)))$, and $Y = L_2(J; L_2(\partial G; L_2(\Omega)))$. The aim of this section is to deduce the regularity properties of the function

$$\zeta(t, x, \omega) := \sum_{k=1}^{\infty} b_k(t, x) X_k(t, \omega) \tag{2.15}$$

where $(X_i)_{i \in \mathbb{N}}$ are entirely independent processes defined on the probability space $(\Omega, \mathcal{F}, \mathbb{P})$ and satisfying Hypothesis (ϕ) (see page 37). The scalar functions $b_i \in L_2(J; L_2(\partial G))$, $i \in \mathbb{N}$, are supposed to be deterministic.

Turning to spatial regularity, we furnish a sufficient and necessary conditions on the multiplier $b := (b_i)_{i \in \mathbb{N}}$, so that the boundary disturbance ζ affiliates to the space

$$Y_s := L_2(J; {}_0W_2^s(\partial G; L_2(\Omega))), \qquad s \geq 0. \tag{2.16}$$

Note that Y_0 is isometrically isomorphic to the basic space Y.

Theorem 2.22. *Let* $s \geq 0$, $G \subset \mathbb{R}^N$ *be a domain with* $C^{[s]+1}$*-boundary and* ζ *given by (2.15).*

(i) *Suppose* $(X_n)_{n \in \mathbb{N}}$ *is a sequence of mutually independent processes with a unique spectral density* ϕ *subject to Hypothesis* (ϕ)*. Then*

$$b \in L_{2,\frac{\gamma-1}{2}}(J; {}_0W_2^s(\partial G; \ell_2)) \quad \Longrightarrow \quad \zeta \in L_2(J; {}_0W_2^s(\partial G; L_2(\Omega))).$$

(ii) *Suppose* $(X_n)_{n \in \mathbb{N}}$ *is a sequence of mutually independent processes with a unique spectral density* ϕ *subject to Hypothesis* (ϕ_0)*. Then*

$$\zeta \in L_2(J; {}_0W_2^s(\partial G; L_2(\Omega))) \quad \Longrightarrow \quad b \in L_{2,\frac{\gamma_0-1+\theta}{2}}(J; {}_0W_2^s(\partial G; \ell_2)).$$

Proof. Without loss of generality we set $0 < s < 1$. Starting with claim (i), the presence of Theorem 2.11(i) and a straight forward computation gives

$$\|\zeta\|_{Y_s} = \left\| \left(\mathbb{E} \left| \sum_{k=1}^\infty b_k(t,x) X_k(t) \right|^2 \right)^{\frac{1}{2}} \right\|_{L_2(J; {}_0W_2^s(\partial G))}$$

$$= \left\| \left(\mathbb{E} \left[\sum_{k=1}^\infty |b_k(t,x) X_k(t)|^2 + \sum_{k \neq l} b_k(t,x) b_l(t,x) X_k(t) X_l(t) \right] \right)^{\frac{1}{2}} \right\|_{L_2(J; {}_0W_2^s(\partial G))}$$

and with the aid of the independence of X_i and X_j for $i \neq j$ we proceed with

$$\|\zeta\|_{Y_s} = \left\| \left(\sum_{k=1}^\infty |b_k(t,x)|^2 \mathbb{E}|X_k(t)|^2 \right)^{\frac{1}{2}} \right\|_{L_2(J; {}_0W_2^s(\partial G))}$$

$$\leq c \left\| \left(\sum_{k=1}^\infty |t^{\frac{\gamma-1}{2}} b_k(t,x)|^2 \right)^{\frac{1}{2}} \right\|_{L_2(J; {}_0W_2^s(\partial G))} = c\|b\|_{L_{2,\frac{\gamma-1}{2}}(J; {}_0W_2^s(\partial G; \ell_2))}.$$

2.4. DETERMINISTIC MULTIPLIERS

Turning to (ii) we recall that is suffices to prove the claim for $J = [0,1]$ and obtain in similar fashion to (i)

$$\|\zeta\|_{Y_s} \geq c_0 \|b\|_{L_{2,\frac{2\varrho-1+\theta}{2}}(J;_0 W_2^s(\partial G;\ell_2))}.$$

Our next aim is to deduce conditions on b, so that the boundary disturbance ζ admits some time regularity. To this end we provide some technical tools with the subsequent lemmata.

Lemma 2.23. *Suppose X is subject to Hypotheses (ϕ) with $|\lambda|^\gamma \phi(\lambda) \equiv \text{const}$ and let $b : J \to \mathbb{R}$ a deterministic function. Then there is a constant $c > 0$, so that*

$$\mathbb{E}\left[b(t)X(t) - b(s)X(s)\right]^2$$
$$= c\left[\left(b(t)t^{\frac{\gamma-1}{2}} - b(s)s^{\frac{\gamma-1}{2}}\right)^2 + |b(t)b(s)|\left(|t-s|^{\gamma-1} - \left[t^{\frac{\gamma-1}{2}} - s^{\frac{\gamma-1}{2}}\right]^2\right)\right]$$

holds for all $s, t \in J$.

Proof. The claim can be shown directly with the aid of Theorem 2.11(i).

$$\mathbb{E}\left[b(t)X(t) - b(s)X(s)\right]^2 = \mathbb{E}\left[b^2(t)X^2(t) + b^2(s)X^2(s) - 2b(t)b(s)X(t)X(s)\right]$$
$$= b^2(t)\mathbb{E}[X(t)]^2 + b^2(s)\mathbb{E}[X(s)]^2 - 2b(t)b(s)\mathbb{E}[X(t)X(s)]$$
$$= c\left[b^2(t)t^{\gamma-1} + b^2(s)s^{\gamma-1} - b(t)b(s)\left(t^{\gamma-1} + s^{\gamma-1} - |t-s|^{\gamma-1}\right)\right].$$

We then proceed with the elementary manipulations

$$\mathbb{E}\left[b(t)X(t) - b(s)X(s)\right]^2$$
$$= c\left[b^2(t)t^{\gamma-1} + b^2(s)s^{\gamma-1} + b(t)b(s)|t-s|^{\gamma-1} - b(t)b(s)t^{\gamma-1} - b(t)b(s)s^{\gamma-1}\right]$$
$$= c\left[\left(b(t)t^{\frac{\gamma-1}{2}} - b(s)s^{\frac{\gamma-1}{2}}\right)^2 + b(t)b(s)\left(|t-s|^{\gamma-1} - \left[t^{\frac{\gamma-1}{2}} - s^{\frac{\gamma-1}{2}}\right]^2\right)\right]$$

and the proof is complete. □

Lemma 2.24. *Suppose X is subject to Hypotheses (ϕ) and let $b : J \to \mathbb{R}$ a deterministic function. Then there is a constant $c > 0$, so that*

$$\mathbb{E}\left[b(t)X(t) - b(s)X(s)\right]^2 \leq c\left[b^2(t)|t-s|^{\gamma-1} + |b(t) - b(s)|^2 s^{\gamma-1}\right]$$

holds for all $s, t \in J$.

Proof. The claim can be shown directly with the aid of Theorem 2.11(i).

$$\mathbb{E}\left[b(t)X(t) - b(s)X(s)\right]^2 = \mathbb{E}\left[b(t)(X(t) - X(s)) + X(s)(b(t) - b(s))\right]^2$$
$$\leq 2\mathbb{E}[b(t)(X(t) - X(s))]^2 + 2\mathbb{E}[X(s)(b(t) - b(s))]^2$$
$$\leq c\left[b^2(t)|t-s|^{\gamma-1} + |b(t) - b(s)|^2 s^{\gamma-1}\right],$$

which completes the proof. □

Theorem 2.25. *Let $G \subset \mathbb{R}^N$ be a domain with boundary of class C^1 and ζ given by (2.15).*

(i) *Suppose $(X_n)_{n \in \mathbb{N}}$ is a sequence of mutually independent processes with a unique spectral density ϕ subject to Hypothesis (ϕ), so that $|\lambda|^\gamma \phi(\lambda) \equiv \text{const}$ and let $0 \leq 2\sigma < \gamma - 1$.*

$$b \in {}_0W^\sigma_{2,\frac{\gamma-1}{2}}(J; L_2(\partial G; \ell_2)) \implies \zeta \in {}_0W^\sigma_2(J; L_2(\partial G; L_2(\Omega))).$$

(ii) *Let the assumptions of (i) be valid and assume further that for all $s, t \in J$ and $x \in \partial G$ it is $(b(t,x)|b(s,x))_{\ell_2} \geq 0$, then*

$$\zeta \in {}_0W^\sigma_2(J; L_2(\partial G; L_2(\Omega))) \implies b \in {}_0W^\sigma_{2,\frac{\gamma-1}{2}}(J; L_2(\partial G; \ell_2)).$$

(iii) *Suppose $(X_n)_{n \in \mathbb{N}}$ is a sequence of mutually independent processes with a unique spectral density ϕ subject to Hypothesis (ϕ) and let $0 \leq 2\sigma < \gamma - 1$. Then*

$$b \in {}_0W^\sigma_2(J; L_2(\partial G; \ell_2)) \implies \zeta \in {}_0W^\sigma_2(J; L_2(\partial G; L_2(\Omega))).$$

Proof. Regarding assertions (i) and (iii), it is due to Lemma 1.1 and Theorem 2.22 with $s = 0$, that $\zeta \in Y$. So it suffices to compare the relevant semi-norms. In the sequel we denote by $[\cdot]_\sigma$ the semi-norm of the space $W^\sigma_2(J; L_2(\partial G; L_2(\Omega)))$ and by $\mathrm{d}\eta$ the surface measure of ∂G. Let moreover be ζ_m the m-th partial sum of ζ, that is

$$\zeta_m(t,x) = \sum_{k=1}^m b_k(t,x) X_k(t).$$

2.4. DETERMINISTIC MULTIPLIERS

Then

$$[\zeta_m]_\sigma^2 = \int_J \int_J \int_{\partial G} \frac{\mathbb{E}[\zeta_m(t,x) - \zeta_m(s,x)]^2}{|t-s|^{1+2\sigma}} \mathrm{d}\eta(x)\mathrm{d}s\mathrm{d}t$$

$$= \int_J \int_J \int_{\partial G} \frac{\sum_{k=1}^m \mathbb{E}\left[b_k(t,x)X_k(t) - b_k(s,x)X_k(s)\right]^2}{|t-s|^{1+2\sigma}} \mathrm{d}\eta(x)\mathrm{d}s\mathrm{d}t. \quad (2.17)$$

With view on (i), Lemma 2.23 yields

$$[\zeta_m]_\sigma^2 = c\int_J \int_J \int_{\partial G} \frac{\sum_{k=1}^m (b_k(t,x)t^{\frac{\gamma-1}{2}} - b_k(s,x)s^{\frac{\gamma-1}{2}})^2}{|t-s|^{1+2\sigma}} \mathrm{d}\eta(x)\mathrm{d}s\mathrm{d}t+$$

$$+ c\int_J \int_J \int_{\partial G} \frac{\sum_{k=1}^m b_k(t,x)b_k(s,x)\left[|t-s|^{\gamma-1} - (t^{\frac{\gamma-1}{2}} - s^{\frac{\gamma-1}{2}})^2\right]}{|t-s|^{1+2\sigma}} \mathrm{d}\eta(x)\mathrm{d}s\mathrm{d}t. \quad (2.18)$$

Let us now study the second term of the right hand-side of (2.18) separately. It is due to $|t-s|^{\gamma-1} \geq (t^{\frac{\gamma-1}{2}} - s^{\frac{\gamma-1}{2}})^2$ for all $s,t \in J$ (to see this multiply with $s^{1-\gamma}$ and substitute $z = t/s$), that

$$\int_J \int_J \int_{\partial G} \frac{\sum_{k=1}^m b_k(t,x)b_k(s,x)\left[|t-s|^{\gamma-1} - (t^{\frac{\gamma-1}{2}} - s^{\frac{\gamma-1}{2}})^2\right]}{|t-s|^{1+2\sigma}} \mathrm{d}\eta(x)\mathrm{d}s\mathrm{d}t$$

$$\leq \int_J \int_J \int_{\partial G} \frac{\sum_{k=1}^m |b_k(t,x)b_k(s,x)||t-s|^{\gamma-1}}{|t-s|^{1+2\sigma}} \mathrm{d}\eta(x)\mathrm{d}s\mathrm{d}t$$

$$\leq \frac{1}{2}\int_J \int_J \int_{\partial G} \frac{\sum_{k=1}^m [b_k^2(t,x) + b_k^2(s,x)]}{|t-s|^{2+2\sigma-\gamma}} \mathrm{d}\eta(x)\mathrm{d}s\mathrm{d}t$$

$$= \int_J \int_J \int_{\partial G} \frac{\sum_{k=1}^m b_k^2(t,x)}{|t-s|^{2+2\sigma-\gamma}} \mathrm{d}\eta(x)\mathrm{d}s\mathrm{d}t.$$

Passing to the limit $m \to \infty$ forces the existence of constants $c_1, c_2 > 0$ (in the sequel generic) so that

$$[\zeta]_\sigma^2 \leq c_1[b]^2_{{}_0W^\sigma_{2,\frac{\gamma-1}{2}}(J;L_2(\partial G;\ell_2))} + c_2 \int_J \int_J \int_{\partial G} \frac{\sum_{k=1}^\infty b_k^2(t,x)}{|t-s|^{2+2\sigma-\gamma}} \mathrm{d}\eta(x)\mathrm{d}s\mathrm{d}t$$

$$= c_1[b]^2_{{}_0W^\sigma_{2,\frac{\gamma-1}{2}}(J;L_2(\partial G;\ell_2))} + c_2 \int_0^T \int_0^t \frac{\|b(t,\cdot)\|^2_{L_2(\partial G;\ell_2)}}{|t-s|^{2+2\sigma-\gamma}} \mathrm{d}s\mathrm{d}t \quad (2.19)$$

$$= c_1[b]^2_{{}_0W^\sigma_{2,\frac{\gamma-1}{2}}(J;L_2(\partial G;\ell_2))} + \frac{c_2}{\gamma-1-2\sigma}\int_0^T \|t^{\frac{\gamma-1}{2}-\sigma}b(t,\cdot)\|^2_{L_2(\partial G;\ell_2)}\mathrm{d}t$$

$$= c_1[b]^2_{{}_0W^\sigma_{2,\frac{\gamma-1}{2}}(J;L_2(\partial G;\ell_2))} + \frac{c_2}{\gamma-1-2\sigma}\|b\|^2_{L_{2,\frac{\gamma-1}{2}-\sigma}(J;L_2(\partial G;\ell_2))}$$

and assertion (i) is established by Lemma 1.1. Turning (ii) we stress that it suffices to prove the claim for $J = [0,1]$ and that Theorem 2.22 yields $b \in L_{2,\frac{\gamma-1}{2}}(J; L_2(\partial G; \ell_2))$. Note that $(b(s,x)|b(t,x))_{\ell_2} \geq 0$ implies the existence of a number $M > 0$ such that for every $m > M$ it is $\sum_{n=1}^m b_n(s,x)b_n(t,x) \geq 0$. Choosing $m > M$ we estimate (2.18) by

$$[\zeta_m]_\sigma^2 = c \int_J \int_J \int_{\partial G} \frac{\sum_{k=1}^m (b_k(t,x)t^{\frac{\gamma-1}{2}} - b_k(s,x)s^{\frac{\gamma-1}{2}})^2}{|t-s|^{1+2\sigma}} d\eta(x)dsdt +$$

$$+ c \int_J \int_J \int_{\partial G} \frac{\sum_{k=1}^m b_k(t,x)b_k(s,x)\left[|t-s|^{\gamma-1} - (t^{\frac{\gamma-1}{2}} - s^{\frac{\gamma-1}{2}})^2\right]}{|t-s|^{1+2\sigma}} d\eta(x)dsdt$$

$$\geq c \int_J \int_J \int_{\partial G} \frac{\sum_{k=1}^m (b_k(t,x)t^{\frac{\gamma-1}{2}} - b_k(s,x)s^{\frac{\gamma-1}{2}})^2}{|t-s|^{1+2\sigma}} d\eta(x)dsdt.$$

Passing to the limit $m \to \infty$ enforces (ii). Let us conclude with the proof of assertion (iii). To this end we may employ Lemma 2.24 to estimate (2.17) by

$$[\zeta_m]_\sigma^2 \leq c \int_J \int_J \int_{\partial G} \frac{\sum_{k=1}^m b_k^2(t,x)}{|t-s|^{2+2\sigma-\gamma}} d\eta(x)dsdt +$$

$$+ c \int_J \int_J \int_{\partial G} \frac{\sum_{k=1}^m |b_k(t,x) - b_k(s,x)|^2 s^{\gamma-1}}{|t-s|^{1+2\sigma}} d\eta(x)dsdt. \quad (2.20)$$

The first term from the right hand-side of (2.20) can be treated as presented in (2.19). The second term from the right hand-side of (2.20) can be handled as

$$\int_J \int_J \int_{\partial G} \frac{\sum_{k=1}^m |b_k(t,x) - b_k(s,x)|^2 s^{\gamma-1}}{|t-s|^{1+2\sigma}} d\eta(x)dsdt$$

$$\leq T^{\gamma-1} \int_J \int_J \int_{\partial G} \frac{\sum_{k=1}^m |b_k(t,x) - b_k(s,x)|^2}{|t-s|^{1+2\sigma}} d\eta(x)dsdt.$$

Passing to the limit $m \to \infty$ yields the existence of constants $c_3, c_4, c_5 > 0$ which may depend on γ, T or σ, so that

$$[\zeta_m]_\sigma^2 \leq c_3 \|b\|^2_{L_{2,\frac{\gamma-1}{2}-\sigma}(J;L_2(\partial G;\ell_2))} + c_4 [b]^2_{\circ W_2^\sigma(J;L_2(\partial G;\ell_2))} \leq c_5 [b]^2_{\circ W_2^\sigma(J;L_2(\partial G;\ell_2))},$$

where the last estimate is verified by Lemma 1.1 and the remark before. □

2.5 Stochastic integration

In what follows let X be a process with stationary increments, so that X has the spectral density ϕ and $X(t) \in L_2(\Omega)$ for every $t \in \mathbb{R}$, if not indicated otherwise. Furthermore we presume, that $X(0) = 0$ a.s. and $\phi(\lambda) > 0$ almost everywhere.

2.5.1 The real-valued case

We will start to study the stochastic integral

$$\mathcal{I}_X(f) := \int_{\mathbb{R}} f(\tau) \mathrm{d}X(\tau), \tag{2.21}$$

where the integrand $f : \mathbb{R} \to \mathbb{R}$ is supposed to be deterministic. If one wants to define this integral as a Riemann-Stieltjes integral, then the class of functions f for which (2.21) is well-defined is rather limited since then f needs to be a continuous function of bounded variation. So we need a different idea in order to define the integral (2.21) for a wider class of functions f. This is the idea of an integral of Itô-type (see [56]). Thus, if f is a step function given by

$$f(t) = \sum_{i=-n}^{n} f_i \chi_{[t_i, t_{i+1})}(t), \tag{2.22}$$

where $t_0 = 0$, we define (2.21) to be

$$\mathcal{I}_X(f) = \sum_{i=-n}^{n} f_i [X(t_{i+1}) - X(t_i)].$$

Obviously, $\mathcal{I}_X(af + bg) = a\mathcal{I}_X(f) + b\mathcal{I}_X(g)$ for any $a, b \in \mathbb{R}$ and step functions f and g. Our aim is to construct a preferably large class of deterministic integrands f so that $\mathcal{I}_X(f)$ is a well-defined random variable with finite second moment.

Remark 2.26. It is readily seen from the construction of $\mathcal{I}_X(f)$ that

1. if X is centered, then so is $\mathcal{I}_X(f)$;

2. if X is Gaussian distributed, then so is $\mathcal{I}_X(f)$.

For the sake of completeness we recall, in the proposition below, how to construct classes of integrands $\mathcal{C}$ for integrals of the form (2.21). This is a generalized version of [86, Proposition 2.1] in this sense, that it is formulated not exclusively for integrals with respect to a fractional Brownian motion, but with respect to a process X with stationary increments featuring the spectral density ϕ.

Proposition 2.27. *Suppose that $\mathcal{C}$ is a set of deterministic functions defined on $\mathbb{R}$ such that*

(a) $\mathcal{C}$ is an inner product space with an inner product $(f \mid g)_\mathcal{C}$, for $f, g \in \mathcal{C}$,

(b) $\mathcal{E} \subset \mathcal{C}$ and $(f \mid g)_\mathcal{C} = (\mathscr{I}_X(f) \mid \mathscr{I}_X(g))_{L_2(\Omega)}$, for $f, g \in \mathcal{E}$,

(c) the set $\mathcal{E}$ is dense in $\mathcal{C}$.

Then there is an isometry between the space $\mathcal{C}$ and a linear subspace of

$$\mathrm{Sp}(X) := \{Y \in L_2(\Omega) : \|\mathscr{I}_X(f_n) - Y\|_{L_2(\Omega)} \to 0, \text{ for some } (f_n) \subset \mathcal{E}\}.$$

which is an extension of the map $f \mapsto \mathscr{I}_X(f)$, for $f \in \mathcal{E}$.

Proof. Let $f \in \mathcal{C}$. By (c), there is a sequence $(f_n) \subset \mathcal{E}$ such that $f_n \to f$ in $\mathcal{C}$. In particular, (f_n) is a Cauchy sequence in $\mathcal{C}$ and hence, by (b), $(\mathscr{I}_X(f_n))$ is a Cauchy sequence in $L_2(\Omega)$. Since the space $L_2(\Omega)$ is complete, there is an element $\mathscr{I}_X(f) \in L_2(\Omega)$ such that

$$\mathscr{I}_X(f) = \lim_{n \to \infty} \mathscr{I}_X(f_n),$$

in the $L_2(\Omega)$ sense. Moreover, since $(\mathscr{I}_X(f_n)) \subset \mathrm{Sp}(X)$ and $\mathrm{Sp}(X)$ is a closed subset of $L_2(\Omega)$, we obtain that $\mathscr{I}_X(f) \in \mathrm{Sp}(X)$. We can thus define the map $\mathscr{I}_X$ from the space $\mathcal{C}$ into the space $\mathrm{Sp}(X)$. Observe, that this definition does not depend on an approximating sequence (f_n). This construction of $\mathscr{I}_X$ and (b) imply that, for $f, g \in \mathcal{C}$,

$$(f \mid g)_\mathcal{C} = (\mathscr{I}_X(f) \mid \mathscr{I}_X(g))_{L_2(\Omega)},$$

and, since the map $\mathscr{I}_X$ is linear, we conclude that $\mathscr{I}_X$ is, in fact, an isometry between the space $\mathcal{C}$ and a linear subspace of $\mathrm{Sp}(X)$. □

2.5. STOCHASTIC INTEGRATION

In the sequel we will denote the isometry map $\mathscr{I}_X$ obtained in the proof above also by

$$\mathscr{I}_X(f) = \int_{\mathbb{R}} f(\tau) \mathrm{d}X(\tau), \tag{2.23}$$

for $f \in \mathcal{C}$, and the right hand-side of (2.23) is called the integral on the real line $\mathbb{R}$ of f with respect to the process X.

Recall the weighted homogeneous Bessel potential space $\dot{\mathrm{H}}_2^\phi(\mathbb{R})$, introduced in Section 1.2.7, with the inner product

$$(f \mid g)_{\dot{\mathrm{H}}_2^\phi(\mathbb{R})} := \int_{\mathbb{R}} \mathcal{F}f(\lambda) \overline{\mathcal{F}g(\lambda)} \lambda^2 \phi(\lambda) \mathrm{d}\lambda.$$

Recall also that by Lemma 1.2 the inner product $(\cdot \mid \cdot)_{\dot{\mathrm{H}}_2^\phi(\mathbb{R})}$ is well-defined and real-valued if the function ϕ is even and almost everywhere positive at least. The next lemma provides an approximation property of distributions belonging to the spaces $\dot{\mathrm{H}}_2^\phi(\mathbb{R})$ and is leaned on Pipiras & Taqqu [86, Lemma 5.1]. It generalizes the results for functions in $\dot{\mathrm{H}}_2^s(\mathbb{R})$, where $|s| < \frac{1}{2}$, to distributions in $\dot{\mathrm{H}}_2^\phi(\mathbb{R})$. However, the proof's strategy remains the same.

Lemma 2.28. *Let the function ϕ satisfying condition (2.12). If $f \in \dot{\mathrm{H}}_2^\phi(\mathbb{R})$, then there is a sequence of elementary functions ψ_n such that*

$$\|\mathcal{F}f - \mathcal{F}\psi_n\|^2_{L_2(\lambda^2 \phi)} = \int_{\mathbb{R}} |\mathcal{F}f(\lambda) - \mathcal{F}\psi_n(\lambda)|^2 \lambda^2 \phi(\lambda) \mathrm{d}\lambda \to 0, \quad \text{as} \quad n \to \infty.$$

Proof. Since for $x \in \mathbb{R}$ it is

$$f(x) = \frac{1}{2}(f(x) + f(-x)) + \frac{1}{2}(f(x) - f(-x))$$

we may prove the lemma in two cases: (1) f is an even function and, (2) f is an odd function.

Case 1: If f is an even function, then $\mathcal{F}f$ is real-valued and $\mathcal{F}f(\lambda) = \mathcal{F}f(-\lambda)$. To prove the claim, we show that, for arbitrary small $\varepsilon > 0$, there is an elementary function ψ such that $\|\mathcal{F}f - \mathcal{F}\psi\|_{L_2(\lambda^2 \phi)} < \varepsilon$. We will provide this approximation in several steps. As a first step, we approximate $\mathcal{F}f$ by simple functions. For any $\varepsilon > 0$

there is a simple function
$$\mathcal{F}g(\lambda) = \sum_{j=1}^{k} g_j \chi_{G_j}(\lambda),$$
where $g_j \in \mathbb{R}$ and $G_j \in \mathcal{B}(\mathbb{R})$, such that
$$\|\mathcal{F}f - \mathcal{F}g\|_{L_2(\lambda^2 \phi)} < \varepsilon.$$

Since $\mathcal{F}f(\lambda) = \mathcal{F}f(-\lambda)$, we can take the sets G_j to be symmetric around the origin $\lambda = 0$. As a second step, observe that, for a symmetric (around the origin) set G and any $\varepsilon > 0$, there is a function
$$\mathcal{F}h(\lambda) = \sum_{n=1}^{m} h_n \chi_{[-H_n, H_n]}(\lambda),$$
with $h_n \in \mathbb{R}$ and $H_n > 0$ such that
$$\|\chi_G - \mathcal{F}h\|_{L_2(\lambda^2 \phi)} < \varepsilon.$$

It is therefore enough to show, that the function $\chi_{[-1,1]}(\lambda)$ can be approximated in $L_2(\lambda^2 \phi)$ by the Fourier transform of an elementary function. In other words, that for any $\varepsilon > 0$ there is an elementary function ψ such that
$$\|\chi_{[-1,1]} - \mathcal{F}\psi\|_{L_2(\lambda^2 \phi)} < \varepsilon.$$

To construct ψ, observe first that
$$\int_{\mathbb{R}} |\chi_{[-1,1]}(\lambda) - \mathcal{F}\psi(\lambda)|^2 \lambda^2 \phi(\lambda) d\lambda = \int_{\mathbb{R}} |\lambda \chi_{[-1,1]}(\lambda) - \lambda \mathcal{F}\psi(\lambda)|^2 \phi(\lambda) d\lambda$$
and by (2.12) the measure $\phi(\lambda)$ is finite around $\lambda = \infty$. The remaining part of the proof is, except of a notations, identically to the proof of [86, Lemma 5.1]. The idea is to truncate the range of $\lambda \chi_{[-1,1]}(\lambda)$, perform a periodic extension and observe that its truncated Fourier series is of the form $\lambda \mathcal{F}\psi(\lambda)$, where $\mathcal{F}\psi$ is the (continuous) Fourier transform of an elementary function. We thus construct ψ as follows. First, choose $k > 1$ so that
$$\int_{\mathbb{R}} \phi(\lambda) \chi_{\{|\lambda|>k\}}(\lambda) d\lambda < \frac{\varepsilon^2}{2}.$$
Now, let U be the function which equals $\lambda \chi_{[-1,1]}(\lambda)$ on $[-k, k]$ and is periodically extended to $\lambda \in \mathbb{R}$. It has the Fourier series $\sum_{n=-\infty}^{\infty} u_n e^{-\frac{i\pi n \lambda}{k}}$, which converges to U

2.5. STOCHASTIC INTEGRATION

everywhere on $[-k, k]$ except at the points $\lambda = \pm 1$, where U is discontinuous. Moreover, the partial sum

$$U_m(\lambda) = \sum_{n=-m}^{m} u_n e^{-\frac{i\pi n \lambda}{k}}$$

can be written as

$$U_m(\lambda) = \frac{1}{k} \int_{-k}^{k} U(\lambda - \xi) D_m\left(\frac{\pi \xi}{k}\right) d\xi,$$

where

$$D_m(\xi) = \frac{\sin(m + \frac{1}{2})\xi}{2\sin(\frac{\xi}{2})}, \quad \xi \in \mathbb{R},$$

is the well-known Dirichlet kernel. The proof of [86, Lemma 5.1] contains the verification of the following properties of the partial sum U_m:

(i) $\sup_m \sup_\lambda |U_m(\lambda)| \leq c$,

(ii) $\sup_m |U_m(\lambda)| \leq c|\lambda|$, for λ small enough,

where the above constants are not necessarily equal the same. By (i) and (ii), the dominated convergence theorem implies that

$$\int_{\{|\lambda| \leq k\}} |\lambda \chi_{[-1,1]}(\lambda) - U_m(\lambda)|^2 \phi(\lambda) d\lambda \to 0$$

as $m \to \infty$. In particular, there is an integer M such that

$$\int_{\{|\lambda| \leq k\}} |\lambda \chi_{[-1,1]}(\lambda) - U_M(\lambda)|^2 \phi(\lambda) d\lambda < \frac{\varepsilon^2}{2}.$$

Since $U(0) = 0$ and $U(-x) = -U(x)$, we have that

$$u_n = \frac{1}{k} \int_{-k}^{k} U(x) e^{-\frac{i\pi n \lambda}{k}} d\lambda = \frac{2i}{k} \int_{0}^{k} U(\lambda) \sin\left(\frac{\pi n \lambda}{k}\right) d\lambda = -ia_n,$$

where $a_n \in \mathbb{R}$, $n \geq 1$. Hence $u_0 = 0$ and $u_n = ia_n$ for $n \leq -1$. Thus

$$U_M(\lambda) = \sum_{n=1}^{M} (-ia_n) \left[e^{\frac{i\pi n \lambda}{k}} - e^{-\frac{i\pi n \lambda}{k}} \right].$$

Since

$$\mathcal{F}\chi_{[-\pi n/k, \pi n/k]}(\lambda) = \frac{e^{\frac{i\pi n \lambda}{k}} - e^{-\frac{i\pi n \lambda}{k}}}{i\lambda},$$

$\lambda^{-1}U_M(\lambda)$ is the Fourier transform of the elementary function

$$\psi = \sum_{n=1}^{M} a_n \chi_{[-\pi n/k, \pi n/k]}(\lambda).$$

We thus obtain the required approximation because

$$\|\chi_{[-1,1]} - \mathcal{F}\psi\|_{L_2(\lambda^2\phi)}^2 = \int_{\mathbb{R}} |\lambda\chi_{[-1,1]}(\lambda) - U_M(\lambda)|^2 \phi(\lambda) d\lambda$$
$$\leq \int_{\{|\lambda|\leq k\}} |\lambda\chi_{[-1,1]}(\lambda) - U_M(\lambda)|^2 \phi(\lambda) d\lambda + \int_{\{|\lambda|>k\}} \phi(\lambda) d\lambda < \varepsilon^2.$$

Case 2: If f is an odd function, then $\mathcal{F}f = i\,\mathrm{Im}\,\mathcal{F}f$ and $\mathrm{Im}\,\mathcal{F}f(-\lambda) = -\,\mathrm{Im}\,\mathcal{F}f(\lambda)$. By the same arguments as in the previous case, it is enough to show that the function $i(\chi_{[0,1]}(\lambda) - \chi_{[-1,0]}(\lambda))$ can be approximated by the Fourier transform of an elementary function. Equivalently, for arbitrary small $\varepsilon > 0$, we have to find an elementary function ψ such that

$$\|(\chi_{[0,1]} - \chi[-1,0]) - i\mathcal{F}\psi\|_{L_2(\lambda^2\phi)} < \varepsilon.$$

The proof is similar to the previous case and we only outline it. Fix k as in the Case 1 and let V be the function which equals

$$\lambda(\chi_{[0,1]}(\lambda) - \chi_{[-1,0]}(\lambda)) = |\lambda|\chi_{[-1,1]}(\lambda)$$

on $[-k, k]$ and is periodically extended to $\lambda \in \mathbb{R}$. Its truncated Fourier series

$$V_m(\lambda) = \sum_{n=-m}^{m} v_n e^{-\frac{i\pi n\lambda}{k}}$$

converges to V everywhere on $[-k, k]$ except at the points $\lambda = \pm 1$. It is not enough here to focus on $V_m(\lambda)$ for small λ because $V_m(0) \neq 0$. Therefore, instead of dealing with $V_m(\lambda)$, we will consider $V_m(\lambda) - V_m(0)$. This function also converges to $V(\lambda)$ almost everywhere and one can show that $\sup_m \sup \lambda |V_m(\lambda) - V_m(0)| \leq c$ and $\sup_m |V_m(\lambda) - V_m(0)| \leq c|\lambda|$, for λ small enough. Moreover,

$$V_m(\lambda) - V_m(0) = \sum_{n=1}^{m} b_n(e^{\frac{i\pi n\lambda}{k}} + e^{-\frac{i\pi n\lambda}{k}} - 2),$$

for some $b_n \in \mathbb{R}$, and hence $\lambda^{-1}(V_m(\lambda) - V_m(0)) = i\mathcal{F}\psi_m$, where ψ_m is the elementary function given by

$$\psi_m = \sum_{n=1}^{m} b_n(\chi_{[0,\pi n/k]} - \chi_{[-\pi n/k, 0]}).$$

2.5. STOCHASTIC INTEGRATION

The conclusion follows as in Case 1. □

Lemma 2.29. Let $X = \{X(t)\}_{t \in \mathbb{R}} \subset L_2(\Omega)$ be a process with stationary increments having the spectral density ϕ. Then we have for $t, s \in \mathbb{R}$ and any $h \geq 0$

$$(X(t+h) - X(t) \mid X(s+h) - X(s))_{L_2(\Omega)} = \left(\chi_{[t,t+h]} \mid \chi_{[s,s+h]}\right)_{\dot{H}_2^\phi(\mathbb{R})}.$$

Proof. In view of identities (2.3) and (2.11) we obtain

$$\begin{aligned}
&(X(t+h) - X(t) \mid X(s+h) - X(s))_{L_2(\Omega)} \\
&= D_2(t+h, s+h) + D_2(t,s) - D_2(t+h, s) - D_2(t, s+h) \\
&= \frac{1}{2}[D(|t-s+h|) - 2D(|t-s|) + D(|t-s-h|)] \\
&= 2\int_0^\infty \{2\cos[(t-s)\lambda] - \cos[(t-s-h)\lambda] - \cos[(t-s+h)\lambda]\}\phi(\lambda)\mathrm{d}\lambda \\
&= \operatorname{Re}\left[\int_{\mathbb{R}} \{2e^{-i(s-t)\lambda} - e^{-i(s-t+h)\lambda} - e^{-i(s-t-h)\lambda}\}\phi(\lambda)\mathrm{d}\lambda\right] \\
&= \operatorname{Re}\left[\int_{\mathbb{R}} \frac{e^{-i\lambda s} - e^{-i\lambda(s+h)}}{i\lambda} \cdot \overline{\left(\frac{e^{-i\lambda t} - e^{-i\lambda(t+h)}}{i\lambda}\right)} \cdot \lambda^2 \phi(\lambda) \mathrm{d}\lambda\right] \\
&= \operatorname{Re}\left[\int_{\mathbb{R}} \mathcal{F}\chi_{[s,s+h]}(\lambda)\overline{\mathcal{F}\chi_{[t,t+h]}(\lambda)}\lambda^2\phi(\lambda)\mathrm{d}\lambda\right] \\
&= \left(\chi_{[s,s+h]} \mid \chi_{[t,t+h]}\right)_{\dot{H}_2^\phi(\mathbb{R})}
\end{aligned}$$

since by Lemma 1.2 the inner product in $\dot{H}_2^\phi(\mathbb{R})$ is real-valued as soon as ϕ is even. □

We are now in the position to formulate the main result for the stochastic integration with respect to random processes with stationary increments and spectral density. Note that the subsequent theorem also allocates an isometry of Itô-type.

Theorem 2.30. Let $X = \{X(t)\}_{t\in\mathbb{R}} \subset L_2(\Omega)$ be a process with stationary increments having the spectral density ϕ. Then for $f, g \in \dot{H}_2^\phi(\mathbb{R})$ it is

$$\mathbb{E}\left[\left(\int_\mathbb{R} f(\tau)\mathrm{d}X(\tau)\right)\left(\int_\mathbb{R} g(\tau)\mathrm{d}X(\tau)\right)\right] = (f \mid g)_{\dot{H}_2^\phi(\mathbb{R})}.$$

In particular, for integrands $f \in \dot{H}_2^\phi(\mathbb{R})$ the integral $\mathscr{I}_X(f)$ given by (2.21) is a well-defined random variable with

$$\mathbb{E}[\mathscr{I}_X(f)]^2 = \|f\|^2_{\dot{H}_2^\phi(\mathbb{R})}.$$

Proof. In view of Lemma 2.28, Proposition 2.27 yields that it suffices to prove the claim for step functions. For this purpose let $f, g \in \mathcal{E} \subset \dot{H}_2^\phi(\mathbb{R})$, that is f, g is of the form (2.22). With the aid of Lemma 2.29 we verify

$$\begin{aligned}(\mathscr{I}_X(f) \mid \mathscr{I}_X(g))_{L_2(\Omega)} &= \left(\sum_{j=-n}^n f_j[X(t_{j+1}) - X(t_j)] \mid \sum_{k=-n}^n g_k[X(t_{k+1}) - X(t_k)]\right)_{L_2(\Omega)} \\ &= \sum_{j=-n}^n \sum_{k=-n}^n f_j g_k \left(X(t_{j+1}) - X(t_j) \mid X(t_{k+1}) - X(t_k)\right)_{L_2(\Omega)} \\ &= \sum_{j=-n}^n \sum_{k=-n}^n f_j g_k \left(\chi_{[t_j,t_{j+1})} \mid \chi_{[t_k,t_{k+1})}\right)_{\dot{H}_2^\phi(\mathbb{R})} \\ &= \left(\sum_{j=-n}^n f_j \chi_{[t_j,t_{j+1})} \mid \sum_{k=-n}^n g_k \chi_{[t_k,t_{k+1})}\right)_{\dot{H}_2^\phi(\mathbb{R})} \\ &= (f \mid g)_{\dot{H}_2^\phi(\mathbb{R})},\end{aligned}$$

which completes the proof. $\square$

We now turn our attention to a vector-valued process $\mathcal{X}$ which is in some sense generated by a mutually independent sequence of processes $(X_n)_{n\in\mathbb{N}}$ having a unique spectral density.

2.5.2 The vector-valued case

Let $\mathcal{H}$ be a separable Hilbert space. We want to deduce properties of functions $R: \mathbb{R} \to \mathcal{B}(\mathcal{H})$ such that the integral

$$\int_{\mathbb{R}} R(t)\mathrm{d}(Q^{1/2}\mathcal{X})(t). \tag{2.24}$$

is well defined. The process $Q^{1/2}\mathcal{X}$ is supposed to satisfy

Hypothesis (X). The operator Q belongs to $\mathscr{L}_1(\mathcal{H})$ is self-adjoint, positive definite and is diagonal with respect to the orthonormal basis $(e_n)_{n \in \mathbb{N}}$ of $\mathcal{H}$, i.e. $Qe_n = \nu_n e_n$ and $\nu_n > 0$ for all $n \in \mathbb{N}$. $\mathcal{X} := \{\mathcal{X}(t)\}_{t \in \mathbb{R}}$ is of the form

$$(\mathcal{X}(t)|x) = \sum_{n=0}^{\infty} X_n(t)(e_n|x), \quad t \in \mathbb{R}, \quad x \in \mathcal{H}, \tag{2.25}$$

where X_n are mutually independent, real-valued and centered processes defined on the probability space $(\Omega, \mathcal{F}, \mathbb{P})$. Moreover, for every $n \in \mathbb{N}$ the process X_n features the unique spectral density ϕ.

As we will see, $\mathcal{X}$ (as in (2.25)) is not a well defined $\mathcal{H}$-valued random variable. However, due to $\mathcal{X}(t) : \Omega \to \mathcal{H}_{Q^{-1/2}}$, where $\mathcal{H}_{Q^{-1/2}}$ is the completion of $\mathcal{H}$ with respect to the norm $|x|^2_{Q^{-1/2}} := |Q^{-1/2}x|_{\mathcal{H}}$, $x \in \mathcal{H}$, the process $Q^{1/2}\mathcal{X}$ converges in $L_2(\Omega; \mathcal{H})$. Note that $Q^{1/2}$ is well-defined and belongs to $\mathscr{L}_2(\mathcal{H})$ and further that the operator $-Q$ is dissipative, which by general results (e.g. Lumer & Phillips [72, Theorem 2.1]) means that Q is invertible.

Let us deduce the distributional properties of the process $Q^{1/2}\mathcal{X}$, where the covariance operator Q and the process $\mathcal{X}$ satisfy Hypothesis (X). It is obvious that $Q^{1/2}\mathcal{X}$ is a process, so it remains to calculate the mean value and the covariance.

$$\mathbb{E}\left(Q^{1/2}\mathcal{X}(t) \mid x\right) = \mathbb{E}\left(\sum_{n=1}^{\infty} \sqrt{\nu_n} X_n(t)e_n \mid x\right) = \sum_{n=1}^{\infty} \sqrt{\nu_n} \mathbb{E}(X_n(t))(e_n \mid x) = 0,$$

for every $x \in \mathcal{H}$ and $t \in \mathbb{R}$. Regarding the covariance we have for all $s, t \in \mathbb{R}$

$$\begin{aligned}(\operatorname{Cov}[Q^{1/2}\mathcal{X}(t), Q^{1/2}\mathcal{X}(s)]e_m \mid e_n) &= \mathbb{E}\left[(Q^{1/2}\mathcal{X}(t) \mid e_m)\left(Q^{1/2}\mathcal{X}(s) \mid e_n\right)\right] \\ &= \delta_{mn}\sqrt{\nu_m\nu_n}\mathbb{E}\left[X_m(t)X_n(s)\right].\end{aligned}$$

Thus it is meaningful to define a vector-valued process of a certain spectral type as

Definition 2.31 (Vector-valued processes of spectral type ϕ). *Let Q and $\mathcal{X}$ be subject to Hypothesis (X), then we call the process $Q^{1/2}\mathcal{X} = \{Q^{1/2}\mathcal{X}(t)\}_{t \in \mathbb{R}}$ defined on $(\Omega, \mathcal{F}, \mathbb{P})$ a $\mathcal{H}$-valued process of spectral type ϕ.*

By the definition of a stochastic integral it is

$$\int_{\mathbb{R}} R(t) \mathrm{d}(Q^{1/2}\mathcal{X})(t) := \sum_{n=1}^{\infty} \int_{\mathbb{R}} R(t) Q^{1/2} e_n \mathrm{d}X_n(t).$$

Our next goal is to calculate the covariance operator. For every $x, y \in \mathcal{H}$ it is

$$\mathbb{E}\left[\left(\int_{\mathbb{R}} R(t)\mathrm{d}(Q^{1/2}\mathcal{X})(t) \mid x\right)_{\mathcal{H}} \left(\int_{\mathbb{R}} R(t)\mathrm{d}(Q^{1/2}\mathcal{X})(t) \mid y\right)_{\mathcal{H}}\right]$$

$$= \mathbb{E}\left[\sum_{k=1}^{\infty}\int_{\mathbb{R}} \left(R(t)Q^{1/2}e_k \mid x\right)_{\mathcal{H}} \mathrm{d}X_k(t) \sum_{l=1}^{\infty}\int_{\mathbb{R}} \left(R(t)Q^{1/2}e_l \mid y\right)_{\mathcal{H}} \mathrm{d}X_l(t)\right]$$

$$= \mathbb{E}\left[\sum_{k=l}\int_{\mathbb{R}} \left(R(t)Q^{1/2}e_k \mid x\right)_{\mathcal{H}} \mathrm{d}X_k(t) \int_{\mathbb{R}} \left(R(t)Q^{1/2}e_l \mid y\right)_{\mathcal{H}} \mathrm{d}X_l(t)\right]$$

$$+ 2\mathbb{E}\left[\sum_{k<l}\int_{\mathbb{R}} \left(R(t)Q^{1/2}e_k \mid x\right)_{\mathcal{H}} \mathrm{d}X_k(t) \int_{\mathbb{R}} \left(R(t)Q^{1/2}e_l \mid y\right)_{\mathcal{H}} \mathrm{d}X_l(t)\right]$$

and with the aid of the independence of X_k and X_l for $k \neq l$ we proceed in view of Theorem 2.30 with

$$\mathbb{E}\left[\left(\int_{\mathbb{R}} R(t)\mathrm{d}(Q^{1/2}\mathcal{X})(t) \mid x\right)_{\mathcal{H}} \left(\int_{\mathbb{R}} R(t)\mathrm{d}(Q^{1/2}\mathcal{X})(t) \mid y\right)_{\mathcal{H}}\right]$$

$$= \sum_{n=1}^{\infty} \mathbb{E}\left[\left(\int_{\mathbb{R}} \left(R(t)Q^{1/2}e_n \mid x\right)_{\mathcal{H}} \mathrm{d}X_n(t)\right)\left(\int_{\mathbb{R}} \left(R(t)Q^{1/2}e_n \mid y\right)_{\mathcal{H}} \mathrm{d}X_n(t)\right)\right]$$

$$= \sum_{n=1}^{\infty} \left((R(\cdot)Q^{1/2}e_n \mid x)_{\mathcal{H}} \mid (R(\cdot)Q^{1/2}e_n \mid y)_{\mathcal{H}}\right)_{\dot{H}_2^{\phi}(\mathbb{R})}.$$

2.5. STOCHASTIC INTEGRATION

Now we may choose $x = y$ to obtain the variance

$$\mathbb{E}\left(\int_{\mathbb{R}} R(t) \mathrm{d}(Q^{1/2}\mathcal{X})(t) \mid x\right)_{\mathcal{H}}^2 = \sum_{n=1}^{\infty} \left\|(R(\cdot)Q^{1/2}e_n \mid x)_{\mathcal{H}}\right\|_{\dot{H}_2^\phi(\mathbb{R})}^2$$

for all $x \in \mathcal{H}$. Hence, letting $(h_n)_{n \in \mathbb{N}}$ an arbitrary orthonormal basis in $\mathcal{H}$, then Parseval's equation yields

$$\mathbb{E}\left|\int_{\mathbb{R}} R(t) \mathrm{d}(Q^{1/2}\mathcal{X})(t)\right|_{\mathcal{H}}^2 = \sum_{k=1}^{\infty} \mathbb{E}\left(\int_{\mathbb{R}} R(t) \mathrm{d}(Q^{1/2}\mathcal{X})(t) \mid h_k\right)_{\mathcal{H}}^2$$
$$= \sum_{k=1}^{\infty} \sum_{n=1}^{\infty} \left\|(R(\cdot)Q^{1/2}e_n \mid h_k)_{\mathcal{H}}\right\|_{\dot{H}_2^\phi(\mathbb{R})}^2.$$

Note, that $R \in \dot{H}_2^\phi(\mathbb{R}; \mathcal{B}(\mathcal{H}))$ implies $RQ^{1/2} \in \dot{H}_2^\phi(\mathbb{R}; \mathscr{L}_2(\mathcal{H}))$ as well as $(RQ^{1/2}e_n \mid x)_{\mathcal{H}} \in \dot{H}_2^\phi(\mathbb{R})$ for all $x \in \mathcal{H}$ and for all $n \in \mathbb{N}$. Resuming, we deduced the following identity.

Theorem 2.32. Let $Q^{1/2}\mathcal{X}$ be an $\mathcal{H}$-valued process of spectral type ϕ. Let further $R : \mathbb{R} \to \mathcal{B}(\mathcal{H})$ and $(h_n)_{n \in \mathbb{N}}$ an orthonormal basis in $\mathcal{H}$. Then the identity

$$\mathbb{E}\left|\int_{\mathbb{R}} R(t) \mathrm{d}(Q^{1/2}\mathcal{X})(t)\right|_{\mathcal{H}}^2 = \sum_{k=1}^{\infty} \sum_{n=1}^{\infty} \left\|(R(\cdot)Q^{1/2}e_n \mid h_k)_{\mathcal{H}}\right\|_{\dot{H}_2^\phi(\mathbb{R})}^2$$

holds and the left hand-side is independent from the choice of the basis $(h_n)_{n \in \mathbb{N}}$. In particular the stochastic integral (2.24) is well defined, if $R \in \dot{H}_2^\phi(\mathbb{R}; \mathcal{B}(\mathcal{H}))$.

Suppose now, that there is a second orthonormal system $(g_n)_{n \in \mathbb{N}}$ in $\mathcal{H}$ and scalar functions $r(\cdot, \cdot) : \mathbb{R} \times \mathbb{C} \to \mathbb{R}$, so that the operator $R(t)$ decomposes into

$$R(t)x = \sum_{n=1}^{\infty} r(t, \mu_n)(x \mid g_n)g_n, \quad t \geq 0, \quad x \in \mathcal{H}.$$

Note that the operator valued function $R : \mathbb{R} \to \mathcal{B}(\mathcal{H})$ particularly admits this property if $R(t)$ is self-adjoint and the resolvent set $\rho(R(t))$ is compact for all $t \in \mathbb{R}$. We have

$$\int_{\mathbb{R}} R(t)\mathrm{d}(Q^{1/2}\mathcal{X})(t) = \sum_{n=1}^{\infty} \int_{\mathbb{R}} R(t)Q^{1/2}e_n \mathrm{d}X_n(t)$$

$$= \sum_{n=1}^{\infty} \sqrt{\nu_n} \int_{\mathbb{R}} R(t)e_n \mathrm{d}X_n(t)$$

$$= \sum_{n=1}^{\infty} \sqrt{\nu_n} \int_{\mathbb{R}} \sum_{k=1}^{\infty} (e_n \mid g_k) r(t, \mu_k) g_k \mathrm{d}X_n(t)$$

$$= \sum_{n=1}^{\infty} \sqrt{\nu_n} \int_{\mathbb{R}} \sum_{k,l} (e_n \mid g_k) r(t, \mu_k) (g_k \mid e_l) e_l \mathrm{d}X_n(t)$$

$$= \sum_{n,k,l} \sqrt{\nu_n} (e_n \mid g_k)(g_k \mid e_l) \int_{\mathbb{R}} r(t, \mu_k) \mathrm{d}X_n(t) e_l$$

and Theorem 2.30 together with the elementary identity $(x \mid y) = \sum_j (x \mid e_j)(y \mid e_j)$ for $x, y \in \mathcal{H}$, yields

$$\mathbb{E} \left| \int_{\mathbb{R}} R(t)\mathrm{d}(Q^{1/2}\mathcal{X})(t) \right|_{\mathcal{H}}^2$$

$$= \sum_{n,k,l,m} \nu_n (e_n \mid g_k)(g_k \mid e_l)(e_n \mid g_m)(g_m \mid e_l) \left(r(\cdot, \mu_k) \mid r(\cdot, \mu_m) \right)_{\dot{H}_2^\phi(\mathbb{R})}$$

$$= \sum_{n,k,m} \nu_n (e_n \mid g_k)(e_n \mid g_m) \left(r(\cdot, \mu_k) \mid r(\cdot, \mu_m) \right)_{\dot{H}_2^\phi(\mathbb{R})} \left[\sum_l (g_k \mid e_l)(g_m \mid e_l) \right]$$

$$= \sum_{n,k} \nu_n (e_n \mid g_k)^2 \left(r(\cdot, \mu_k) \mid r(\cdot, \mu_k) \right)_{\dot{H}_2^\phi(\mathbb{R})}$$

$$= \sum_k \|r(\cdot, \mu_k)\|_{\dot{H}_2^\phi(\mathbb{R})}^2 \left[\sum_n (Q^{1/2}e_n \mid g_k)^2 \right]$$

$$= \sum_k \|r(\cdot, \mu_k)\|_{\dot{H}_2^\phi(\mathbb{R})}^2 \left[\sum_n (e_n \mid Q^{1/2}g_k)^2 \right]$$

and we may employ Parseval's equation $|x|_{\mathcal{H}}^2 = \sum_j (x, e_j)^2$ for $x \in \mathcal{H}$ to conclude

$$\mathbb{E} \left| \int_{\mathbb{R}} R(t)\mathrm{d}(Q^{1/2}\mathcal{X})(t) \right|_{\mathcal{H}}^2 = \sum_{k=1}^{\infty} |Q^{1/2}g_k|_{\mathcal{H}}^2 \|r(\cdot, \mu_k)\|_{\dot{H}_2^\phi(\mathbb{R})}^2.$$

Let us introduce a notation of a shifted, time inverted and truncated function f supported on $\mathbb{R}$ in virtue of

$$f^{\langle t \rangle}(\tau) := f(t-\tau)\chi_{(-\infty,t]}(\tau) = \begin{cases} f(t-\tau) & : -\infty < \tau \leq t; \\ 0 & : \tau > t. \end{cases} \quad (2.26)$$

2.5. STOCHASTIC INTEGRATION

For brevity we write $r_n(\tau) := r(\tau, \mu_n)$. The latter observations lead us to

$$\mathbb{E}\left|\int_{\mathbb{R}}(R^{\langle t\rangle}(\tau) - R^{\langle s\rangle}(\tau))\mathrm{d}(Q^{1/2}\mathcal{X})(\tau)\right|_{\mathcal{H}}^2 = \sum_{k=1}^{\infty}|Q^{1/2}g_k|_{\mathcal{H}}^2\|r_k^{\langle t\rangle} - r_k^{\langle s\rangle}\|_{\dot{\mathrm{H}}_2^\phi(\mathbb{R})}^2.$$

Observe that due to

$$(\mathcal{F}f^{\langle t\rangle})(\xi) = \int_{\mathbb{R}} f(t-\tau)\chi_{(-\infty,t]}(\tau)e^{-i\xi\tau}\mathrm{d}\tau$$
$$= \int_{\mathbb{R}} f(-s)\chi_{(-\infty,0]}(s)e^{-i\xi(s+t)}\mathrm{d}s = e^{-i\xi t}(\mathcal{F}f^{\langle 0\rangle})(\xi)$$

we find for all $t \in \mathbb{R}$

$$\|f^{\langle t\rangle}\|_{\dot{\mathrm{H}}_2^\sigma(\mathbb{R})} = \|f^{\langle 0\rangle}\|_{\dot{\mathrm{H}}_2^\sigma(\mathbb{R})} = \|f\|_{\dot{\mathrm{H}}_2^\sigma(\mathbb{R}_+)}.$$

Next, deduce that in the case where ϕ satisfies Hypothesis (ϕ) (see page 37) we have

$$\dot{\mathrm{H}}_2^{\frac{2-\gamma}{2}}(\mathbb{R}) \hookrightarrow \dot{\mathrm{H}}_2^\phi(\mathbb{R}),$$

and therefrom we obtain by designating with $c > 0$ a generic constant

$$\|r_k^{\langle t\rangle} - r_k^{\langle s\rangle}\|_{\dot{\mathrm{H}}_2^\phi(\mathbb{R})}^2 \leq c\|r_k^{\langle t\rangle} - r_k^{\langle s\rangle}\|_{\dot{\mathrm{H}}_2^{\frac{2-\gamma}{2}}(\mathbb{R})}^2$$
$$= c\left\|\partial^{\frac{2-\gamma}{2}}(r_k^{\langle t\rangle} - r_k^{\langle s\rangle})\right\|_{L_2(\mathbb{R})}^2$$
$$= c\int_{\mathbb{R}}\left|(\partial^{\frac{2-\gamma}{2}}r_k^{\langle t\rangle})(\tau) - (\partial^{\frac{2-\gamma}{2}}r_k^{\langle s\rangle})(\tau)\right|^2\mathrm{d}\tau$$
$$= c\int_{\mathbb{R}}\left|(\partial^{\frac{2-\gamma}{2}}r_k^{\langle 0\rangle})(\tau - t) - (\partial^{\frac{2-\gamma}{2}}r_k^{\langle 0\rangle})(\tau - s)\right|^2\mathrm{d}\tau$$
$$= c\left\|(\partial^{\frac{2-\gamma}{2}}r_k^{\langle 0\rangle})(s - t + \cdot) - (\partial^{\frac{2-\gamma}{2}}r_k^{\langle 0\rangle})(\cdot)\right\|_{L_2(\mathbb{R})}^2$$
$$\leq c\left\|\partial^{\frac{2-\gamma}{2}}r_k^{\langle 0\rangle}\right\|_{B_{2,\infty}^\theta(\mathbb{R})}^2 |t-s|^{2\theta},$$

where $0 \leq \theta < \frac{\gamma-1}{2}$ and $B_{2,\infty}^\theta(\mathbb{R})$ denotes a Besov space with the equivalent norm

$$\|f\|_{B_{2,\infty}^\theta(\mathbb{R})} = \left[\|f\|_{L_2(\mathbb{R})}^2 + \sup_{h\neq 0}\int_{\mathbb{R}}\frac{|f(y+h)-f(y)|^2}{|h|^{2\theta}}\mathrm{d}y\right]^{1/2}$$

and cf. [106, Section 2.3.2] to verify

$$\mathrm{H}_2^\theta(\mathbb{R}) \hookrightarrow B_{2,\infty}^\theta(\mathbb{R}).$$

Finally, with the apparent relation

$$\|f\|_{\dot{\mathrm{H}}_2^\sigma(\mathbb{R})} + \|f\|_{\dot{\mathrm{H}}_2^{\theta+\sigma}(\mathbb{R})} = \|\partial^\sigma f\|_{\mathrm{H}_2^\theta(\mathbb{R})}, \qquad -\tfrac{1}{2} < \sigma < \tfrac{1}{2},$$

we derived the estimate

$$\|r_k^{(t)} - r_k^{(s)}\|^2_{\dot{H}_2^\phi(\mathbb{R})} \leq \left[\|r_k\|_{\dot{H}_2^{\frac{2-\gamma}{2}}(\mathbb{R}_+)} + \|r_k\|_{\dot{H}_2^{\theta+\frac{2-\gamma}{2}}(\mathbb{R}_+)}\right]^2 |t-s|^{2\theta}.$$

Let us consider

Hypothesis (X_ϕ). $Q^{1/2}\mathcal{X}$ is a $\mathcal{H}$-valued process of spectral type ϕ and the all elements of the generating sequence $(X_n)_{n\in\mathbb{N}}$ are subject to Hypothesis (ϕ) (see page 37) with the unique density ϕ and the same parameter $1 < \gamma < 3$.

We have then observed

Theorem 2.33. Let $Q^{1/2}\mathcal{X}$ be subject to Hypothesis (X_ϕ). Let further $R : \mathbb{R} \to \mathcal{B}(\mathcal{H})$ and $(g_n)_{n\in\mathbb{N}}$ be an orthonormal basis in $\mathcal{H}$, such that $R(t)$ decomposes into

$$R(t)x = \sum_{n=1}^\infty r(t, \mu_n)(x \mid g_n)g_n, \quad t \in \mathbb{R}, \quad x \in \mathcal{H}.$$

Then there is a constant $c > 0$ such that

$$\mathbb{E}\left|\int_\mathbb{R} R(t)\mathrm{d}(Q^{1/2}\mathcal{X})(t)\right|^2_\mathcal{H} \leq c \sum_{k=1}^\infty |Q^{1/2}g_k|^2_\mathcal{H} \|r(\cdot, \mu_k)\|^2_{\dot{H}_2^{\frac{2-\gamma}{2}}(\mathbb{R})}.$$

By means of the notation (2.26)

$$\mathbb{E}\left|\int_\mathbb{R} R^{(t)}(\tau)\mathrm{d}(Q^{1/2}\mathcal{X})(\tau)\right|^2_\mathcal{H} \leq c \sum_{k=1}^\infty |Q^{1/2}g_k|^2_\mathcal{H} \|r(\cdot, \mu_k)\|^2_{\dot{H}_2^{\frac{2-\gamma}{2}}(\mathbb{R}_+)}$$

holds true for all $t \in \mathbb{R}$. Moreover, for $\theta \in [0, \frac{\gamma-1}{2})$ and $s, t \in \mathbb{R}$ it is

$$\mathbb{E}\left|\int_\mathbb{R} (R^{(t)}(\tau) - R^{(s)}(\tau))\mathrm{d}(Q^{1/2}\mathcal{X})(\tau)\right|^2_\mathcal{H}$$
$$\leq c \sum_{k=1}^\infty |Q^{1/2}g_k|^2_\mathcal{H} \left[\|r_k\|_{\dot{H}_2^{\frac{2-\gamma}{2}}(\mathbb{R}_+)} + \|r_k\|_{\dot{H}_2^{\theta+\frac{2-\gamma}{2}}(\mathbb{R}_+)}\right]^2 |t-s|^{2\theta}.$$

Concerning integrals evaluated on an interval $[0, t_0]$, where $t_0 > 0$, we deduce the following corollary.

Corollary 2.34. *Let $Q^{1/2}\mathcal{X}$ be subject to Hypothesis (X_o). Let further $R : \mathbb{R} \to \mathcal{B}(\mathcal{H})$ and $(g_n)_{n \in \mathbb{N}}$ be an orthonormal basis in $\mathcal{H}$, such that $R(t)$ decomposes into*

$$R(t)x = \sum_{n=1}^{\infty} r(t, \mu_n)(x \mid g_n) g_n, \quad t \in \mathbb{R}, \quad x \in \mathcal{H}$$

and let $0 \leq s \leq t_0$. Then there is a constant $c > 0$ such that

$$\mathbb{E} \left| \int_0^{t_0} R(\tau) \mathrm{d}(Q^{1/2}\mathcal{X})(\tau) \right|_{\mathcal{H}}^2 \leq c \sum_{k=1}^{\infty} |Q^{1/2} g_k|_{\mathcal{H}}^2 \| r(\cdot, \mu_k) \|_{\dot{\mathrm{H}}_2^{\frac{2-\gamma}{2}}([0,t_0])}^2.$$

Moreover, we have

$$\mathbb{E} \left| \int_0^{t_0} R(t_0 - \tau) \mathrm{d}(Q^{1/2}\mathcal{X})(\tau) \right|_{\mathcal{H}}^2 \leq c \sum_{k=1}^{\infty} |Q^{1/2} g_k|_{\mathcal{H}}^2 \| r(\cdot, \mu_k) \|_{\dot{\mathrm{H}}_2^{\frac{2-\gamma}{2}}([0,t_0])}^2$$

and by means of notation (2.26)

$$\mathbb{E} \left| \int_0^{t_0} (R^{\langle t_0 \rangle}(\tau) - R^{\langle s \rangle}(\tau)) \mathrm{d}(Q^{1/2}\mathcal{X})(\tau) \right|_{\mathcal{H}}^2$$
$$\leq c \sum_{k=1}^{\infty} |Q^{1/2} g_k|_{\mathcal{H}}^2 \left[\| r_k \|_{\dot{\mathrm{H}}_2^{\frac{2-\gamma}{2}}([0,t_0])} + \| r_k \|_{\dot{\mathrm{H}}_2^{\theta + \frac{2-\gamma}{2}}([0,t_0])} \right]^2 |t_0 - s|^{2\theta}.$$

Proof. The claim follows by employing Theorem 2.33 to the operators $R\chi_{[0,t_0]}$, $R^{\langle t_0 \rangle} \chi_{[0,t_0]}$ and $(R^{\langle t_0 \rangle} - R^{\langle s \rangle}) \chi_{[0,t_0]}$ respectively. □

2.6 Examples

2.6.1 Centered Poisson processes

In the probabilistic theory of queues, which has many important applications, the so called Poisson process $\mathcal{P}^\nu$ is quite useful. It is one of the most well-known Lévy processes.

Definition 2.35 (Poisson process). *A real-valued process $\mathcal{P}^\nu = \{\mathcal{P}^\nu(t)\}_{t\in\mathbb{R}}$ defined on a probability space $(\Omega, \mathscr{F}, \mathbb{P})$ is called a Poisson process with intensity $\nu > 0$, if*

(i) $\mathbb{P}\{\mathcal{P}^\nu(0) = 0\} = 1$;

(ii) *For $\tau > 0$ the increments $\mathcal{P}^\nu(t+\tau) - \mathcal{P}^\nu(t)$ are Poisson distributed with parameter $\nu\tau$, i.e.*
$$\mathbb{P}\{\mathcal{P}^\nu(t+\tau) - \mathcal{P}^\nu(t) = k\} = \frac{e^{-\nu\tau}(\nu\tau)^k}{k!}, \quad k \in \mathbb{N}_0;$$

(iii) *For $s < t < u < v$ the increments $\mathcal{P}^\nu(t) - \mathcal{P}^\nu(s)$ and $\mathcal{P}^\nu(v) - \mathcal{P}^\nu(u)$ are stochastically independent.*

We shall further assume, that no counted occurrences are simultaneous. As already mentioned Poisson processes are suitable to model counting phenomena, so for instance

- the long-term behavior of the number of web page requests arriving at a server, except for unusual circumstances such as coordinated denial of service attacks or flash crowds;

- the number of telephone calls arriving at a switchboard, or at an automatic phone-switching system;

- the number of photons hitting a photodetector, when lit by a laser source;

- the execution of trades on a stock exchange, as viewed on a tick by tick basis.

It is already seen from Definition 2.35(ii) that the number of occurrences counted in any time interval only depends on the length of the interval which in turn means, that $\mathcal{P}^\nu$ features stationary increments. Observe further from Definition 2.35(ii), that $\mathbb{E}[\mathcal{P}^\nu(t)] = \nu t$ for all $t \in \mathbb{R}$. Thus the process $\mathcal{P}_0^\nu(t) := \mathcal{P}^\nu(t) - \nu t$ is a centered Poisson process with intensity ν. It is the process $\mathcal{P}_0^\nu$ which will be of further interest. It is now readily seen from the stationarity of the increments and Definition 2.35(i) that

$$\mathbb{E}[\mathcal{P}_0^\nu(-\tau)]^2 = \mathbb{E}[\mathcal{P}_0^\nu(t+\tau) - \mathcal{P}_0^\nu(t)]^2 = \mathbb{E}[\mathcal{P}_0^\nu(\tau)]^2,$$

2.6. EXAMPLES

hence, by Definition 2.35(ii), the structure function D of the process $\mathcal{P}_0^\nu$ is of the form $D(\tau) = \nu|\tau|$. Then the function D can also be represented in the form (2.11), where

$$\phi(\lambda) = \frac{\nu}{2\pi|\lambda|^2}. \tag{2.27}$$

Thus the spectrum of $\mathcal{P}_0^\nu$ contains the whole real line $\mathbb{R}$.

Remark 2.36. Note that every centered Lévy process, that is a centered process with stationary and independent (or at least uncorrelated) increments, has a spectral density of the form $\phi(\lambda) = c|\lambda|^{-2}$ with a positive constant $c > 0$, as soon as the density exists. Note further that the following results only require this particular structure of the spectral density (2.27).

It is apparent by Example 2.6, that $\mathcal{P}_0^\nu$ is subject to Hypotheses (ϕ) and (ϕ_0) (see page 37) with $\gamma = \gamma_0 = 2$ and $\theta = 0$. Following Remark 2.5, we outline that $\gamma_0 = 2$ indicates the absence of long-range dependence and $\theta = 0$ signals that centered Poisson processes (or, more generally, centered Lévy processes) are not appropriate to study intermittency effects.

Corollary 2.37. Let $\mathcal{P}_0^\nu$ be a centered Poisson process with intensity $\nu > 0$. Then the following are true.

(i) The trajectories of $\mathcal{P}_0^\nu$ are almost surely nowhere mean-square differentiable.

(ii) Let $T > 0$. If and only if $0 < \sigma < \frac{1}{2}$. Then $\mathcal{P}_0^\nu \in {}_0W_2^\sigma([0,T]; L_2(\Omega))$.

Proof. Assertions (i) and (ii) follow from Theorem 2.18, while (iii) is a consequence of Theorem 2.21. □

Concerning stochastic integration we can capture the well-known result (e.g. Applebaum [11])

Corollary 2.38. *Let $\mathcal{P}_0^\nu$ be a centered Poisson process with intensity $\nu > 0$. Then for all $f, g \in L_2(\mathbb{R})$ we have the isometry*

$$\mathbb{E}\left[\left(\int_\mathbb{R} f(\tau)\mathrm{d}\mathcal{P}_0^\nu(\tau)\right)\left(\int_\mathbb{R} g(\tau)\mathrm{d}\mathcal{P}_0^\nu(\tau)\right)\right] = \frac{\nu}{2\pi}(f \mid g)_{L_2(\mathbb{R})}.$$

Proof. The claim is immediate by Theorem 2.30. □

Focusing our multiplier results from Theorems 2.22 and 2.25, we denote

$$\zeta(t, x, \omega) := \sum_{k=1}^\infty b_k(t, x)\mathcal{P}_{0,k}^\nu(t, \omega),$$

where $(\mathcal{P}_{0,k}^\nu)_{k\in\mathbb{N}}$ are entirely independent centered Poisson processes with intensity $\nu > 0$ and the scalar functions $b_i \in L_2(J; L_2(\partial G))$, $i \in \mathbb{N}$, are supposed to be deterministic. Denoting $b := (b_i)_{i\in\mathbb{N}}$, we directly obtain from Theorems 2.22 and 2.25 the subsequent corollaries.

Corollary 2.39. *Let $s \geq 0$ and $G \subset \mathbb{R}^N$ be a domain with boundary of class $C^{[s]+1}$. Then*

$$b \in L_{2,1/2}(J; {}_0W_2^s(\partial G; \ell_2)) \iff \zeta \in L_2(J; {}_0W_2^s(\partial G; L_2(\Omega))).$$

Corollary 2.40. *Let $G \subset \mathbb{R}^N$ be a domain with boundary of class C^1 and $0 < \sigma < \frac{1}{2}$. Then*

$$b \in {}_0W_{2,1/2}^\sigma(J; L_2(\partial G; \ell_2)) \implies \zeta \in {}_0W_2^\sigma(J; L_2(\partial G; L_2(\Omega))).$$

If, in addition, $(b(s, x) \mid b(t, x))_{\ell_2} \geq 0$ for every $s, t \in J$, $x \in \partial G$, then the converse is also true.

2.6.2 Fractional Brownian motions

The concept of a fractional Brownian motion was first proposed by Mandelbrot & van Ness [76] and is formally the convolution of Wiener increments with a power-law kernel. More abstractly we formulate

2.6. EXAMPLES

Definition 2.41 (Fractional Brownian motion). *A real-valued Gaussian process $B^H = \{B^H(t)\}_{t \in \mathbb{R}}$ defined on a probability space $(\Omega, \mathscr{F}, \mathbb{P})$ is called a fractional Brownian motion with Hurst parameter $0 < H < 1$, if*

  (i) *B^H is centered,*

  (ii) *$\mathbb{E}[B^H(t)B^H(s)] = \frac{c}{2}\left[|t|^{sH} + |s|^{2H} - |t-s|^{2H}\right]$, where $c > 0$ and $t, s \in \mathbb{R}$.*

Observe that in case $H = \frac{1}{2}$ Definition 2.41 coincides with the definition of a Wiener process. Unlike a Wiener process, a fractional Brownian motion with Hurst parameter $H \neq \frac{1}{2}$ is neither a martingale, nor a semi-martingale, nor Markovian.

It follows directly from Definition 2.41, that a fractional Brownian motion is a process with stationary increments according to Definition 2.2. The first condition holds since the expectation operator $\mathbb{E}$ is linear and B^H is centered, because then for all $t, s \in \mathbb{R}$ it is

$$\mathbb{E}[B^H(t) - B^H(s)] = 0 = \mathbb{E}[B^H(t-s) - B^H(0)].$$

The second condition of Definition 2.2 is also satisfied. This can be seen by the very elementary computation

$$\begin{aligned}
D_3(t; u, v) &= \mathbb{E}\left[(B^H(u) - B^H(t))(B^H(v) - B^H(t))\right] \\
&= \mathbb{E}\left[B^H(u)B^H(v) - B^H(u)B^H(t) - B^H(t)B^H(v) + B^H(t)B^H(t)\right] \\
&= \frac{c}{2}\left[|u-t|^{2H} + |v-t|^{2H} - |u-v|^{2H}\right] \\
&= \mathbb{E}\left[B^H(u-t)B^H(v-t)\right] = D_2(u-t, v-t).
\end{aligned}$$

Moreover, Definition 2.41(ii) yields $D(\tau) = c|t|^{2H}$ which can be written in the form (2.11), where

$$\phi(\lambda) = \frac{c_1}{|\lambda|^{2H+1}}, \quad c_1 = \frac{c}{4\int_0^\infty (1-\cos\lambda)\lambda^{-2H-1}d\lambda} = \frac{c\Gamma(2H+1)\sin(\pi H)}{2\pi}. \quad (2.28)$$

Thus the spectrum of B^H incorporates all $\lambda \in \mathbb{R}$ and in the fashion of Example 2.6 one verifies that the process B^H satisfies Hypotheses (ϕ) and (ϕ_0) (see page 37) if and only if $\gamma = \gamma_0 = 2H + 1$ and $\theta = 0$. Following Remark 2.5, we outline that

$\gamma_0 = 2H + 1$ indicates the presence of long-range dependence in all the cases where $H > \frac{1}{2}$ and $\theta = 0$ signals that fractional Brownian motions are not appropriate to study intermittency effects. We summarize the properties of a fractional Brownian motion in the following corollary.

Corollary 2.42. *Let B^H be a fractional Brownian motion with Hurst parameter $H \in (0, 1)$. Then the following are true.*

(i) B^H is mean-square continuous and has continuous paths almost sure.

(ii) The trajectories of B^H are locally Hölder-continuous of any order strictly less then H.

(iii) The trajectories of B^H are almost surely nowhere mean-square differentiable.

(iv) Let $T > 0$, $0 < p < \infty$, $2 \leq q < \infty$ and $0 < \sigma < H$. Then $B^H \in {_0}W_p^\sigma([0,T]; L_q(\Omega))$.

(v) Let $T > 0$, $0 < p < \infty$, $1 < q \leq 2$ and $H \leq \sigma < 1$. Then $B^H \notin {_0}W_p^\sigma([0,T]; L_q(\Omega))$.

Proof. Assertions (i)-(iii) follow from Theorem 2.18, while (iv) and (v) are consequences of Theorem 2.21. □

The paths of B^H get less zigzagged as H goes from 0 to 1. On this basis, one typically classifies fractional Brownian motions into antipersistent (in case $0 < H < \frac{1}{2}$), chaotic (in case $H = \frac{1}{2}$) and persistent (in case $\frac{1}{2} < H < 1$). This can be loosely explained by considering the covariance of two consecutive increments. When $0 < H < \frac{1}{2}$, the increments of B^H tend to have opposite signs. On the other hand, in case $\frac{1}{2} < H < 1$, the correlation of two consecutive increments is strictly positive.

The following figures were generated with the aid of a Wolfram Demonstration Project contributed by R. E. Maeder.[1]

---
[1] See http://demonstrations.wolfram.com/OneDimensionalFractionalBrownianMotion/.

2.6. EXAMPLES

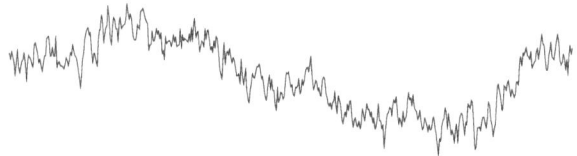

Figure 2.1. *A sample path of a fractional Brownian motion with Hurst parameter* $H = 0.2$.

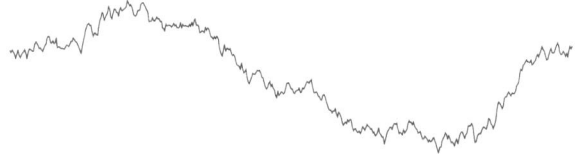

Figure 2.2. *A sample path of a fractional Brownian motion with Hurst parameter* $H = 0.5$.

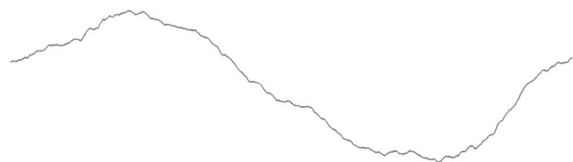

Figure 2.3. *A sample path of fractional Brownian motion with Hurst parameter* $H = 0.9$.

Regarding the fractional dimension of the graphs of it due to Falconer [43, Theorem 16.7] that with probability 1, the Hausdorff and box dimension of the graph $(t, B^H(t))_{0 \leq t \leq 1}$ equal $2 - H$. Thus, if H is close to zero, the process B^H zigzags so much that the dimension of the graph $(t, B^H(t))_{0 \leq t \leq 1}$ is close to the dimension 2 of the unit square. On the other hand B^H is an index-H random field (cf. Angulo et al. [5]) so that the Hausdorff dimension of its image $\{B^H(t) : t \in [0,1]\}$ equals 1 a.s. for every $H \in (0,1)$.

Concerning stochastic integration we can easily reproduce the results of, e.g., Pipiras & Taqqu [86], Sp. & Wilke [104] and Biagini et al. [17].

Corollary 2.43. *Let B^H be a fractional Brownian motion with Hurst parameter $0 < H < 1$. Then for all $f, g \in \dot{H}_2^{\frac{1}{2}-H}(\mathbb{R})$ we have the isometry*

$$\mathbb{E}\left[\left(\int_\mathbb{R} f(\tau)\mathrm{d}B^H(\tau)\right)\left(\int_\mathbb{R} g(\tau)\mathrm{d}B^H(\tau)\right)\right] = c_1 \int_\mathbb{R} (\mathcal{F}f)(\lambda)\overline{(\mathcal{F}g)(\lambda)}|\lambda|^{1-2H}\mathrm{d}\lambda$$

with the constant c_1 from (2.28).

Proof. The claim is immediate by Theorem 2.30. □

Note, that by Plancherel's Theorem (cf. Theorem A.1) our result also covers the well-known Itô-isometry in case $H = \frac{1}{2}$. Focusing our multiplier results from Chapter 2 we denote

$$\zeta(t, x, \omega) := \sum_{k=1}^\infty b_k(t, x) B_k^H(t, \omega),$$

where $(B_k^H)_{k \in \mathbb{N}}$ are entirely independent fractional Brownian motions with Hurst parameter $0 < H < 1$ and the scalar functions $b_i \in L_2(J; L_2(\partial G))$, $i \in \mathbb{N}$, are supposed to be deterministic. Denoting $b := (b_i)_{i \in \mathbb{N}}$, we have in this particular situation

Corollary 2.44. *Let $s \geq 0$ and $G \subset \mathbb{R}^N$ be a domain with boundary of class $C^{[s]+1}$. Then*

$$b \in L_{2,H}(J; {}_0W_2^s(\partial G; \ell_2)) \iff \zeta \in L_2(J; {}_0W_2^s(\partial G; L_2(\Omega))).$$

as a result of Theorem 2.22 and also by Theorem 2.25

Corollary 2.45. *Let $G \subset \mathbb{R}^N$ be a domain with boundary of class C^1 and $0 < \sigma < H$. Then*

$$b \in {}_0W_{2,H}^\sigma(J; L_2(\partial G; \ell_2)) \implies \zeta \in {}_0W_2^\sigma(J; L_2(\partial G; L_2(\Omega))).$$

If, in addition, $(b(s, x)|b(t, x))_{\ell_2} \geq 0$ for every $s, t \in J$, $x \in \partial G$, then the converse is also true.

2.6. EXAMPLES

Interpreting the spectral density (2.28) tempts to differentiate B^H and claim that the fractional noise $\dot{B}^H$ has a spectral density proportional to λ^{1-2H} which suggests that in case $H > \frac{1}{2}$, there is infinite energy at high frequencies and coincides with the known fact that white noise ($H = \frac{1}{2}$) has a flat power spectrum. Thus, for $H \in [\frac{1}{2}, 1)$ fractional white noise (also called $1/f^\alpha$-noise with $0 \leq \alpha = 1 - 2H < 1$) interpolates between white noise ($1/f^0$-noise) and pink noise ($1/f^1$-noise), which is a signal with a frequency spectrum such that the power spectral density is proportional to the reciprocal of the frequency. The following figures were generated with the aid of a Matlab routine contributed by the SAMP group, based at the Department of Engineering Science, Oxford University.[2]

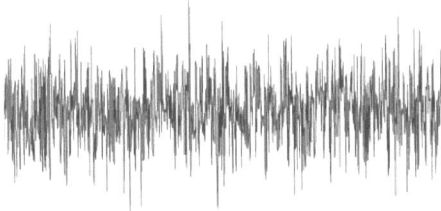

Figure 2.4. Sample of $1/f^0$-noise.

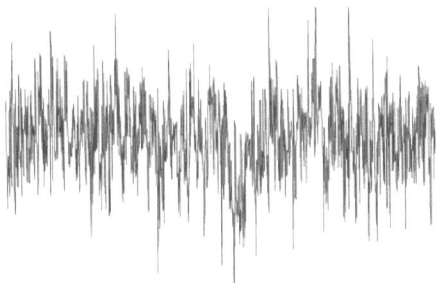

Figure 2.5. Sample of $1/f^{1/2}$-noise.

---

[2] See http://www.eng.ox.ac.uk/samp/powernoise_soft.html.

78 CHAPTER 2. PROCESSES WITH STATIONARY INCREMENTS

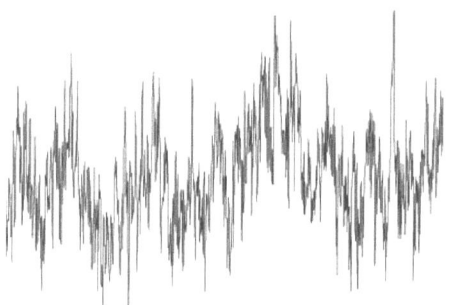

Figure 2.6. *Sample of* $1/f^1$*-noise.*

By having power at all frequencies, the total energy of a $1/f^\alpha$-noise is infinite, so it is apparent that such a signal only exists as a theoretical model. In practice, approximations are used where the spectral density decays rapidly for high frequencies. However, $1/f^\alpha$-noise occurs widely and has applications in a large number of fields, so for instance:

- White noise is used by some emergency sirens, due to its ability to cut through background noise, which makes it easier to locate.

- White noise is used extensively in audio synthesis, typically to recreate percussive instruments such as cymbals which have high noise content in their frequency domain.

- The sound of a waterfall results from the collision of drops among themselves or with the water surface. The smaller the drops the more efficient is air friction, so that small drops are stronger decelerated. At the impact the smaller drops are consequently slower than bigger ones, such that they contribute only faintly high frequency sound fractures. Therefrom the sound of a waterfall is approximately pink noise.

- The human auditory system, which processes frequencies in a roughly logarithmic fashion, does not perceive them with equal sensitivity; signals in the 24-kHz

2.6. EXAMPLES

octave sound loudest, and the loudness of other frequencies drops increasingly. While white noise is de facto equitable loud in every bandwidth, people sense pink noise having this feature rather than white noise.

Getting more and more general we are now accomplish to fractional Riesz-Bessel motions.

2.6.3 Fractional Riesz-Bessel motions

Based on a concept of duality of generalized stochastic processes defined on fractional Sobolev spaces introduced in Ruiz-Medina et al. [96], Anh et al. [6] proved the existence of a class of stochastic processes defined by the equation

$$(I - \Delta)^{\frac{\alpha}{2}}(-\Delta)^{\frac{\beta}{2}} X(x) = \dot{B}^{\frac{1}{2}}(x), \qquad x \in \mathbb{R},$$

where $\dot{B}^{\frac{1}{2}}$ is white noise, or equivalently (in the sense of second order moments) by the spectral density

$$\phi(\lambda) = \frac{1}{|\lambda|^{2\beta}(1+\lambda^2)^{\alpha}}, \qquad (2.29)$$

where $\frac{1}{2} < \beta < \frac{3}{2}$, $\alpha \geq 0$ and $\lambda \in \mathbb{R}$. These processes were named fractional Riesz-Bessel motions, here denoted by $\mathcal{RB}_{\alpha}^{\beta}$. It is a Gaussian process with stationary increments, whose spectral density involves the Fourier transforms of the Riesz kernel and the Bessel kernel. Comparing the spectral densities (2.28) and (2.29) it can be seen, that a fractional Brownian motion B^H is the special case of a fractional Riesz-Bessel motion $\mathcal{RB}_{\alpha}^{\beta}$ for $\beta = H + \frac{1}{2}$ and $\alpha = 0$. Since this is the case we exclude the case $\alpha = 0$ in the sequel.

Following Anh & Nguyen [9] we define

Definition 2.46 (Fractional Riesz-Bessel motion). *A real-valued Gaussian process $\mathcal{RB}_{\alpha}^{\beta} := \{\mathcal{RB}_{\alpha}^{\beta}(t)\}_{t \in \mathbb{R}}$ defined on the probability space $(\Omega, \mathcal{F}, \mathbb{P})$ is called a fractional Riesz-Bessel motion with parameters $\alpha > 0$ and $\frac{1}{2} < \beta < \frac{3}{2}$, if*

(i) $\mathbb{P}\{\mathcal{RB}_{\alpha}^{\beta}(0) = 0\} = 1$,

(ii) $\mathcal{RB}_\alpha^\beta$ has stationary increments,

(iii) $\mathcal{RB}_\alpha^\beta$ has a spectral density of the form (2.29).

It is readily seen, that the spectrum of $\mathcal{RB}_\alpha^\beta$ covers the real line $\mathbb{R}$. Note further, that the spectral density of the stationary increments, which behaves as $|\lambda|^{2-2\beta}$ as $|\lambda| \to 0$, has a singularity at frequency 0, if and only if $\beta > 1$. Note further, that the component $(1+\lambda^2)^{-\alpha}$ in (2.29) indicates the second-order intermittency, i.e. clustering of extreme values (as $|\lambda| \to \infty$) of order $\alpha + \beta$.

Observe with the aid of Example 2.6, that the spectral density (2.29) clearly satisfies Hypothesis (ϕ) (see page 37) with $\gamma \in [2\beta, 2(\alpha+\beta)] \cap [2\beta, 3)$. Moreover (2.29) is also due to Hypothesis (ϕ_0) with $\gamma_0 = 2\beta$ and $\theta = 2\alpha$. Following Remark 2.5, we outline that $\gamma_0 = 2\beta$ indicates the presence of long-range dependence in all the cases where $\beta > 1$ and $\theta = 2\alpha$ signals that fractional Riesz-Bessel motions are appropriate to study intermittency effects, whenever $\alpha > 0$. We summarize the properties of a fractional Riesz-Bessel motion in the following corollary.

Corollary 2.47. Let $\mathcal{RB}_\alpha^\beta$ be a fractional Riesz-Bessel motion with parameters $\frac{1}{2} < \alpha + \beta < \frac{3}{2}$.

(i) For all $\tau \in \mathbb{R}$ the structure function of $\mathcal{RB}_\alpha^\beta$ satisfies

$$\mathbb{E}[\mathcal{RB}_\alpha^\beta(\tau)]^2 \leq \min\left\{c_\phi(\gamma)|\tau|^{\gamma-1} : \gamma \in [2\beta, 2(\alpha+\beta)]\right\},$$

and also

$$\mathbb{E}[\mathcal{RB}_\alpha^\beta(\tau)]^2 \geq c_{\phi_0} \cdot \min\{|\tau|^{2(\alpha+\beta)-1}, |\tau|^{2\beta-1}\},$$

with the constants

$$c_\phi(\gamma) = 2^{4-\gamma}\int_0^\infty \frac{\sin^2(\lambda)}{\lambda^\gamma}d\lambda, \quad c_{\phi_0} = \max\left\{\frac{2^{4-2\beta}}{(1+\lambda_0^2)^\alpha}\int_0^{\lambda_0/2}\frac{\sin^2(\lambda)}{\lambda^{2\beta}}d\lambda : \lambda_0 > 0\right\}.$$

(ii) If $\alpha + \beta > 1$, then $\mathcal{RB}_\alpha^\beta$ is a centered process.

(iii) If $\alpha + \beta > 1$, then $\mathcal{RB}_\alpha^\beta$ is mean-square continuous and has continuous paths a.s.

2.6. EXAMPLES

(iv) If $\alpha + \beta > 1$, then the trajectories of $\mathcal{RB}_\alpha^\beta$ are locally Hölder-continuous of order strictly less then $\alpha + \beta - \frac{1}{2}$.

(v) The trajectories of $\mathcal{RB}_\alpha^\beta$ are almost surely nowhere mean-square differentiable.

(vi) Let $T > 0$ and $0 < \sigma < 1$. If and only if $\sigma < \alpha + \beta - \frac{1}{2}$, then $\mathcal{RB}_\alpha^\beta \in {}_0W_2^\sigma([0,T]; L_2(\Omega))$.

(vii) Let $\alpha + \beta > 1$, $T > 0$, $2 \leq q < \infty$, $0 < p < \infty$ and $0 < \sigma < 1$. If $\sigma < \alpha + \beta - \frac{1}{2}$, then $\mathcal{RB}_\alpha^\beta \in {}_0W_p^\sigma([0,T]; L_q(\Omega))$.

(viii) Let $\alpha + \beta > 1$, $T > 0$, $1 < q \leq 2$, $0 < p < \infty$ and $0 < \sigma < 1$. If $\sigma \geq \alpha + \beta - \frac{1}{2}$, then $\mathcal{RB}_\alpha^\beta \notin {}_0W_p^\sigma([0,T]; L_q(\Omega))$.

Proof. Assertion (i) follows from Theorem 2.11 and Corollary 2.12, (iii)-(v) are due to Theorem 2.18, (ii) is due to Corollary 2.19, while (vi)-(viii) are consequences of Theorem 2.21. □

Figures 2.7 and 2.8 (see below) illustrate the quality of the estimates for the second moments $\mathbb{E}[\mathcal{RB}_\alpha^\beta(t)]^2$ provided by Corollaries 2.12 and 2.13, respectively. The major observance is, that in every case an increasing parameter α decelerates the growth of $\mathbb{E}[\mathcal{RB}_\alpha^\beta(t)]^2$ and widens the shaded region. Although the lower bound seems not to be sharp for small values of α, it is a matter of fact, that it is approached by the concrete values of $\mathbb{E}[\mathcal{RB}_\alpha^\beta(t)]^2$ as t tends to infinity. Conversely, the upper bound is meaningful for large values of α as $t \to \infty$.

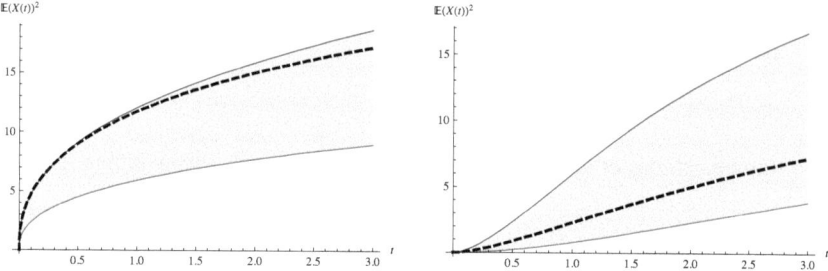

Figure 2.7. The actual values (dashed) and the predicted region of the second moments $\mathbb{E}[X(t)]^2$ with $X = \mathcal{RB}_\alpha^\beta$, where $\alpha = 0.1$ and $\beta = 0.6$ (left) resp. $\alpha = 0.89$ and $\beta = 0.6$ (right).

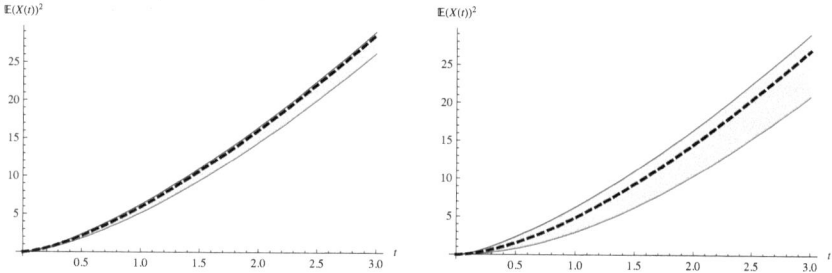

Figure 2.8. The actual values (dashed) and the predicted region of the second moments $\mathbb{E}[X(t)]^2$ with $X = \mathcal{RB}_\alpha^\beta$, where $\alpha = 0.05$ and $\beta = 1.2$ (left) resp. $\alpha = 0.29$ and $\beta = 1.2$ (right).

The result concerning stochastic integration is rather new and reads as

Corollary 2.48. Let $\mathcal{RB}_\alpha^\beta$ be a fractional Riesz-Bessel motion. Then for all $f, g \in \dot{\mathrm{H}}_2^\phi(\mathbb{R})$ with ϕ given by (2.29), we have the isometry

$$\mathbb{E}\left[\left(\int_\mathbb{R} f(\tau)\mathrm{d}\mathcal{RB}_\alpha^\beta(\tau)\right)\left(\int_\mathbb{R} g(\tau)\mathrm{d}\mathcal{RB}_\alpha^\beta(\tau)\right)\right] = \int_\mathbb{R} (\mathcal{F}f)(\lambda)\overline{(\mathcal{F}g)(\lambda)}\frac{|\lambda|^{2-2\beta}}{(1+\lambda^2)^\alpha}\mathrm{d}\lambda.$$

Proof. The claim is immediate by Theorem 2.30. □

2.6. EXAMPLES

Focusing our multiplier results from Chapter 2 we denote

$$\zeta(t,x,\omega) := \sum_{k=1}^{\infty} b_k(t,x) \mathcal{RB}^{\beta}_{\alpha,k}(t,\omega),$$

where $(\mathcal{RB}^{\beta}_{\alpha,k})_{k\in\mathbb{N}}$ are entirely independent fractional Riesz-Bessel motions with parameters $1 < 2(\alpha+\beta) < 3$ and the scalar functions $b_i \in L_2(J; L_2(\partial G))$, $i \in \mathbb{N}$, are supposed to be deterministic. Denoting $b := (b_i)_{i\in\mathbb{N}}$, we have in this particular situation

Corollary 2.49. Let $s \geq 0$ and $G \subset \mathbb{R}^N$ be a domain with boundary of class $C^{[s]+1}$. Then

$$b \in L_{2,\alpha+\beta-\frac{1}{2}}(J; {}_0W_2^s(\partial G; \ell_2)) \iff \zeta \in L_2(J; {}_0W_2^s(\partial G; L_2(\Omega))).$$

as a result of Theorem 2.22 and also by Theorem 2.25

Corollary 2.50. Let $G \subset \mathbb{R}^N$ be a domain with boundary of class C^1 and $0 \leq \sigma < \alpha + \beta - \frac{1}{2}$. Then

$$b \in {}_0W_2^{\sigma}(J; L_2(\partial G; \ell_2)) \implies \zeta \in {}_0W_2^{\sigma}(J; L_2(\partial G; L_2(\Omega))).$$

In order to illustrate the behavior of the paths of $\mathcal{RB}^{\beta}_{\alpha}$ under a variation of the parameter α we present Figures 2.9 – 2.12 below, generated with the aid of a simulation proposed by Anh et al. [1, Section 5.2]. As a turnout it can be seen, that intermittency effects amplifies with an increasing parameter α. In addition, we see a smoothing ef-

84   CHAPTER 2. PROCESSES WITH STATIONARY INCREMENTS

fect as the noise profile becomes somehow "thinner".

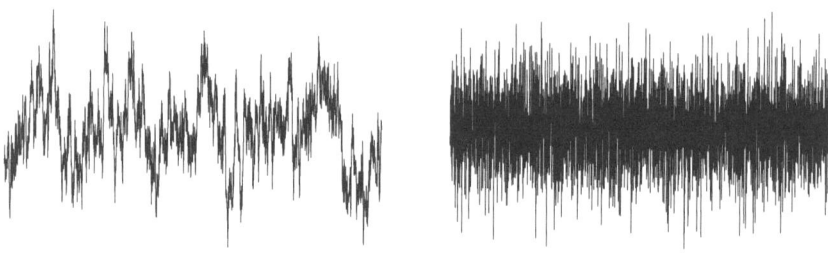

Figure 2.9. *A random path of a fractional Riesz-Bessel motion (left) and corresponding noise profile (right) with parameters* $\beta = 0.7$ *and* $\alpha = 0.1$.

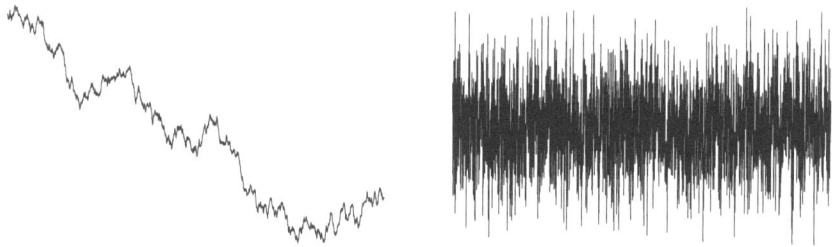

Figure 2.10. *A random path of a fractional Riesz-Bessel motion (left) and corresponding noise profile (right) with parameters* $\beta = 1.05$ *and* $\alpha = 0.1$.

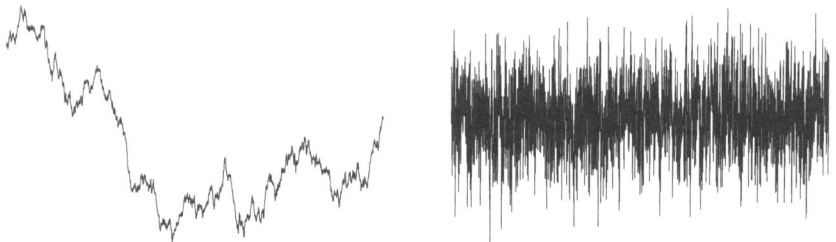

Figure 2.11. *A random path of a fractional Riesz-Bessel motion (left) and corresponding noise profile (right) with parameters* $\beta = 1.05$ *and* $\alpha = 0.44$.

2.6. EXAMPLES

The paths of a fractional Riesz-Bessel motion become less zigzagged as β goes from $\frac{1}{2}$ to $\frac{3}{2}$. Such a smoothing effect can also be seen in Figures 2.1 - 2.3. For the choice $\alpha + \beta > \frac{3}{2}$ a fractional Riesz-Bessel motion is moreover mean-square differentiable (cf. Anh & Nguyen [10, Proposition 9]). This feature is depicted with the subsequent figure.

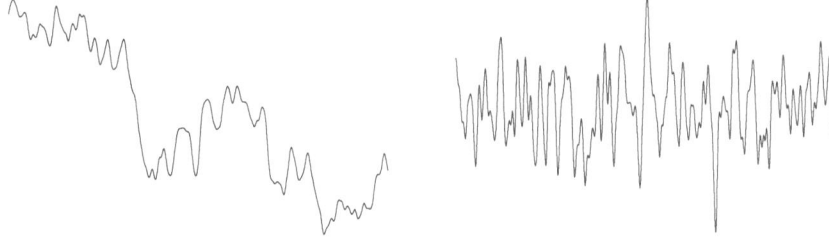

Figure 2.12. *A random path of a fractional Riesz-Bessel motion (left) and corresponding noise profile (right) with parameters $\beta = 1.05$ and $\alpha = 300$.*

Chapter 3

Parabolic Volterra equations

Aim of this chapter is to study different types of parabolic Volterra equations with random disturbances. Our plan is to present the main result first. Throughout this chapter $\mathcal{H}$ is a separable Hilbert space and $Q^{1/2}\mathcal{X}$ is subject to Hypothesis (X_ϕ) (see page 68). Thus we have seen in Section 2.6 that the following results will in particular cover the cases, where the disturbance is modeled to be a centered Lévy process, a fractional Brownian motion B^H with Hurst parameter $H \in (0,1)$, or a fractional Riesz-Bessel motion $\mathcal{RB}_\sigma^\eta$ with parameters $\frac{1}{2} < \eta < \frac{3}{2}$ and $\sigma \geq 0$, such that $\mathcal{RB}_\sigma^\eta$ is centered.

3.1 Main results

Let A be a closed linear densely defined operator in $\mathcal{H}$, and $b \in L_1(\mathbb{R}_+)$ a scalar kernel. Let us consider the problem

$$u(t) + \int_0^t b(t-\tau)Au(\tau)\mathrm{d}\tau = Q^{1/2}\mathcal{X}(t), \qquad t \geq 0 \qquad (3.1)$$

in the Hilbert space $\mathcal{H}$. In particular we recall that Hypothesis (X_ϕ) forces the existence of a sequence $(\nu_n)_{n\in\mathbb{N}} \in \ell_1(\mathbb{R}_+)$ and an orthonormal basis $(e_n)_{n\in\mathbb{N}} \subset \mathcal{H}$, such that $Qe_n = \nu_n e_n$ for every $n \in \mathbb{N}$.

Because problem (3.1) is motivated from applications of linear viscoelastic material behavior, we consider the operator $-A$ to be an elliptic differential operator like the Laplacian, the elasticity operator, or the Stokes operator, together with appropriate boundary conditions (e.g. Prüss [88, Section I.5]). We formulate abstractly

Hypothesis (A). A is an unbounded, self-adjoint, positive definite operator in $\mathcal{H}$ with compact resolvent. Consequently, the eigenvalues μ_n of A form a strictly positive, nondecreasing sequence with $\lim_{n\to\infty}\mu_n = \infty$, the corresponding eigenvectors $(a_n)_{n\in\mathbb{N}} \subset \mathcal{H}$ form an orthonormal basis of $\mathcal{H}$.

Observe, that Hypothesis **(A)** implies the sectoriality of the operator A with angle $\phi_A = 0$ (cf. [34, Section 1]). This observation allows us to define complex powers A^z for arbitrary $z \in \mathbb{C}$; cf. [88, Section 8.1].

The kernel b is supposed to be the antiderivative of a 3-monotone scalar function (see Definition 1.8); more precisely b is subject to

Hypothesis (b): The kernel b is of the form

$$b(t) = b_0 + \int_0^t b_1(\tau)\mathrm{d}\tau, \quad t > 0, \tag{3.2}$$

where $b_0 \geq 0$ and $b_1(t)$ is 3-monotone with $\lim_{t\to\infty} b_1(t) = 0$; in addition,

$$\lim_{t\downarrow 0} \frac{\frac{1}{t}\int_0^t \tau b_1(\tau)\mathrm{d}\tau}{b_0 + \int_0^t -\tau \dot{b}_1(\tau)\mathrm{d}\tau} < \infty. \tag{3.3}$$

In case **(A)** and **(b)** are valid, problem (3.1) is well-posed and parabolic; for kernels subject to (3.2), condition (3.3) is in fact equivalent to parabolicity. Typical examples of kernels arising from the theory of linear viscoelasticity (cf. [88, Section I.5]), which satisfy Hypothesis **(b)** are the material functions of Newtonian fluids ($b_0 > 0$, $b_1 \equiv 0$), Maxwell fluids ($b_0 = 0$, $b_1(t) = \sigma\exp\{-\frac{\sigma t}{\nu}\}$) and of power type materials ($b_0 = 0$, $b_1(t) = g_\alpha(t)$, $\alpha \in (1,2)$). Define

$$\rho := \frac{2}{\pi}\sup\left\{|\arg\widehat{b}(\lambda)| : \operatorname{Re}\lambda > 0\right\},$$

3.1. MAIN RESULTS

then we obtain the subsequent existence and regularity results for the mild solution of (3.1).

Theorem 3.1. Let Hypotheses *(A)*, *(b)* and *(X_ϕ)* are valid.

(i) If $QA^{\frac{1-\gamma}{\rho}} \in \mathscr{L}_1(\mathcal{H})$, then the mild solution u of (3.1) exists and is mean-square continuous on $\mathbb{R}_+$. Moreover, the trajectories of u are continuous on the half-line $\mathbb{R}_+$ almost surely.

(ii) If in addition, there is $\theta \in (0, \frac{\gamma-1}{2})$ such that $QA^{\frac{1-\gamma}{\rho}+\frac{2\theta}{\rho}} \in \mathscr{L}_1(\mathcal{H})$, then the trajectories of u are locally Hölder-continuous of any order strictly less then θ almost surely.

In advantage to [25] we do not require that the eigensystems of the operators A and Q has to coincide, that is if $a_n = e_n$ for all $n \in \mathbb{N}$. If, as in our situation, the eigensystems of A and Q can be arbitrary orthonormal bases of $\mathcal{H}$ it is meaningful to say, that the perturbation $Q^{1/2}\mathcal{X}$ is "system independent". On the other hand we will say, that the perturbation $Q^{1/2}\mathcal{X}$ is "A-synchronized" if the eigensystems of A and Q coincide. Regarding existence and regularity it is easily seen from Theorem 3.1 and Theorem 3.5 (see below) that the results are independent from the choice of the eigensystems.

Remark 3.2. The case $b \equiv \text{const}$ merely corresponds to the stochastic differential equation

$$\begin{cases} \dot{u} + Au = Q^{1/2}\dot{\mathcal{X}}, & t > 0, \\ u(0) = 0. \end{cases}$$

It is then obvious that Theorem 3.1 applies with $\rho = 1$. Moreover, the notions of strong and mild solutions in the sense of Definition 1.3 are equivalent in all the cases where $b \equiv \text{const}$.

Example 3.3. Let $\mathcal{H} = L_2(0, \pi)$ and consider

$$u(t) + (b * Au)(t) = Q^{1/2} B^H(t), \quad t \geq 0, \tag{3.4}$$

where b is due to Hypothesis (b). Set $A = A_0^m$, where $m \in \mathbb{N}$ and $A_0 = -(\mathrm{d}/\mathrm{d}x)^2$ with domain $D(A_0) = \mathrm{H}_2^2(0, \pi) \cap {}_0\mathrm{H}_2^2(0, \pi)$ and let $0 < H < 1$, where further $Qe_n = n^{-\nu} e_n$ for $n \in \mathbb{N}$. Observe that A is due to Hypothesis (A) and possesses the eigenvalues $\mu_k = k^{2m}$ for $k \in \mathbb{N}$. Theorem 3.1 yields that for every ϕ_b-sectorial kernel b with $\phi_b \in [\frac{\pi}{2}, \pi)$ (this corresponds to $\rho \in [1, 2)$) the mild solution u of (3.4) exists and that its trajectories are Hölder-continuous of any order strictly less than H.

An interesting occurs if it is assumed that $Q = I$, i.e. for all $x \in \mathcal{H}$ it is $Qx = x$, so that $Q^{1/2} B^H = B^H$ is only a cylindrical fractional Brownian motion.

Example 3.4. Assume the setting of Example 3.3, but let $Q = I$. Then the mild solution u of (3.4) exists for every ϕ_b-sectorial kernel b with $\phi_b \in [\frac{\pi}{2}, \min\{2\pi m H; \pi\})$. Moreover its trajectories are Hölder-continuous of any order $\theta < H - \frac{2\phi_b}{4\pi m}$. Note that in this case θ depends on the sectoriality-angle ϕ_b and the exponent m. Highly regular kernels b (this corresponds to ϕ_b near π) cause a loss in time regularity, while an increasing exponent m improves Hölderianity.

Let us take up a different viewpoint to Volterra equations with fractional noise. We consider the problems

$$u(t) + \int_0^t g_\alpha(t - \tau) Au(\tau) \mathrm{d}\tau = \int_0^t g_\beta(t - \tau) \mathrm{d}(Q^{1/2} \mathcal{X})(\tau), \quad t \geq 0 \tag{3.5}$$

in the Hilbert space $\mathcal{H}$, where the operator A is subject to Hypothesis (A) and g_κ denotes the Riemann-Liouville kernel; see (1.4).

In case (A) is valid and $0 < \alpha < 2$, problem (3.5) is well-posed and parabolic.

3.1. MAIN RESULTS

Theorem 3.5. *Assume Hypotheses (A) and (X_ϕ) are valid and let $\alpha \in (0,2)$, $\beta > 0$, $\theta \in [0,1]$, such that $\beta \in (\frac{3-\gamma}{2} + \theta, \frac{3-\gamma}{2} + \theta + \alpha)$.*

(i) *If $QA^{\frac{3-2\beta-\gamma}{\alpha}} \in \mathscr{L}_1(\mathcal{H})$, then the mild solution u of (3.5) exists and is mean-square continuous on $\mathbb{R}_+$. Moreover, the trajectories of u are almost surely continuous on $\mathbb{R}_+$.*

(ii) *If $QA^{\frac{3-2\beta-\gamma}{\alpha} + \frac{2\theta}{\alpha}} \in \mathscr{L}_1(\mathcal{H})$, then the trajectories of u are locally Hölder-continuous of any order strictly less then θ almost surely.*

On the first view Theorem 3.1 seems to be a variant of Theorem 3.5 for the case $\beta = 1$ and more general kernels. However, Hypothesis (b) is too stringent as to countenance standard kernels g_α with $\alpha < 1$.

Remark 3.6. Note that

1. if one chooses $\mathcal{X}$ to be a vector-valued centered Lévy process, then the above results hold with $\gamma = 2$.

2. if one chooses $\mathcal{X} = B^H$ to be a vector-valued fractional Brownian motion with Hurst parameter $0 < H < 1$ the above results hold with $\gamma = 2H + 1$.

3. if one chooses $\mathcal{X} = \mathcal{RB}_\sigma^\eta$ to be a vector-valued fractional Riesz-Bessel motion with parameters $\frac{1}{2} < \eta < \frac{3}{2}$ and $\sigma \geq 0$, then the above results hold true for every $\gamma \in [2\eta, 2(\sigma + \eta)] \cap [2\eta, 3)$.

4. in case $\mathcal{X} = B^H$ with $H = \frac{1}{2}$, that is a vector-valued Wiener process, Theorem 3.5 captures the setting of Clément et al. [25, Theorem 4.2]. However, our approach elevates the upper bound for the feasible β from $\frac{1}{2} + \alpha$ to $\frac{1}{2} + \alpha + \theta$. This is due to the fact, that we estimated the scalar kernels r_n in terms of their $\dot{\mathrm{H}}_2^\theta$-norms, instead of their H_2^θ-norms.

Example 3.7. Let $\mathcal{H} = L_2(0,\pi)$, $\mathcal{X} = \mathcal{RB}^\eta_\sigma$, a A-synchronized $\mathcal{H}$-valued fractional Riesz-Bessel motion with parameters $1 < \eta + \sigma < \frac{3}{2}$, $A = A_0^m$, where $A_0 = -(d/dx)^2$ with domain $D(A_0) = H_2^2(0,\pi) \cap {}_0H_2^1(0,\pi)$. It is obvious that A is subject to Hypothesis (A) and it is well-known that the eigenvalues of A are $\mu_k = k^{2m}$ for $k \in \mathbb{N}$. The covariance Q is supposed to be A-synchronized and is given by its spectral decomposition

$$Qx = \sum_{k=1}^\infty \nu_k(x|e_k)e_k,$$

with $(\nu_k)_{k\in\mathbb{N}} \subset (0,1]$ such that $\sum_{k=1}^\infty \nu_k < \infty$. For our example we choose $\nu_k = k^{-l}$, $l > 1$, and we obtain

$$\sum_{k=1}^\infty \nu_k \mu_k^{\frac{3-2\beta-2\eta-2\sigma}{\alpha}} < \infty \iff \beta > \frac{3}{2} - \eta - \sigma - \frac{\alpha(l-1)}{4m};$$

$$\sum_{k=1}^\infty \nu_k \mu_k^{\frac{3-2\beta+2\theta-2\eta-2\sigma}{\alpha}} < \infty \iff \beta > \frac{3}{2} - \eta - \sigma + \theta - \frac{\alpha(l-1)}{4m}.$$

Obviously the latter two series converge for all

$$\frac{3}{2} - \eta - \sigma + \theta < \beta < \frac{3}{2} - \eta - \sigma + \theta + \alpha,$$

hence Theorem 3.5 applies independently from the choice of l and m. Observe that the temporal regularity increases as $\eta + \sigma$ goes from 1 to $\frac{3}{2}$.

3.2 Proof of the main results

This section is devoted to the proof of the chapter's main results. Before doing so we prove some elementary results needed later on.

Lemma 3.8. Let $\mu > 0$, the function b satisfying Hypothesis (b) with $\rho \in [1,2)$ and denote by $s : \mathbb{R}_+ \to \mathbb{R}$ the solution of the problem

$$s(t) + \mu(b*s)(t) = 1, \quad t \geq 0$$

Then the following are true.

3.2. PROOF OF THE MAIN RESULTS

(i) $|s(t)| \leq 1$ for all $t \geq 0$;

(ii) $\|\dot{s}\|_{L_1(\mathbb{R}_+)} \leq c$;

(iii) $\|s\|_{L_1(\mathbb{R}_+)} \leq c\mu^{-\frac{1}{p}}$;

(iv) $\|s\|_{\dot{H}_2^{\theta+\frac{1}{2}-\sigma}(\mathbb{R}_+)} \leq c\mu^{\frac{\theta-\sigma}{p}}$ for $\sigma \in (0,1)$, $\theta \in [0,\sigma)$ and $\mu \geq 1$,

where $c > 0$ denotes a constant which is independent of μ.

Proof. Assertion (i) follows from the proof of [88, Corollary 1.2], while (ii) is contained in [78, Proposition 6] (observe the relation $\dot{s}(t) = -\mu r_\mu(t)$, to connect the notations) and (iii) is proven in [25, Lemma 3.1]. Turning to (iv) we recall that for all real numbers r and $1 \leq p < \infty$ it is trivially $H_p^r(\mathbb{R}_+) \hookrightarrow \dot{H}_p^r(\mathbb{R}_+)$. For now let $\theta = 0$. If $\sigma = \frac{1}{2}$, then by (i) and (iii) we obtain

$$\|s\|_{\dot{H}_2^0(\mathbb{R}_+)} \leq \|s\|_{H_2^0(\mathbb{R}_+)} = \|s\|_{L_2(\mathbb{R}_+)} \leq \|s\|_{L_1(\mathbb{R}_+)}^{\frac{1}{2}} \leq c\mu^{-\frac{1}{2p}}.$$

In case $\sigma < \frac{1}{2}$, we make use of $L_{1/\sigma}(\mathbb{R}_+) \hookrightarrow \dot{H}_2^{\frac{1}{2}-\sigma}(\mathbb{R}_+)$ (cf. [86, Proposition 3.2]). Therewith assertions (i) and (iii) force

$$\|s\|_{\dot{H}_2^{\frac{1}{2}-\sigma}(\mathbb{R}_+)} \leq c\|s\|_{L_{1/\sigma}(\mathbb{R}_+)} \leq c\|s\|_{L_1(\mathbb{R}_+)}^{\sigma} \leq c\mu^{-\frac{\sigma}{p}}.$$

If $\sigma > \frac{1}{2}$, then [106, Theorem 2.4.7] provides

$$[L_1(\mathbb{R}_+); H_q^r(\mathbb{R}_+)]_\delta = H_2^{\frac{1}{2}-\sigma}(\mathbb{R}_+), \quad \text{where} \quad \delta = 1-\sigma, \quad \tau = \frac{1-2\sigma}{2-2\sigma}, \quad q = \frac{1}{\tau}.$$

Consequently, the interpolation inequality yields

$$\|s\|_{\dot{H}_2^{\frac{1}{2}-\sigma}(\mathbb{R}_+)} \leq \|s\|_{H_2^{\frac{1}{2}-\sigma}(\mathbb{R}_+)} \leq c\|s\|_{L_1(\mathbb{R}_+)}^{\sigma}\|s\|_{H_q^\tau(\mathbb{R}_+)}^{1-\sigma}, \quad c > 0.$$

In what follows let $c > 0$ be generic. One may apply [106, Theorem 2.7.1] to verify $H_1^1(\mathbb{R}_+) \hookrightarrow H_q^\tau(\mathbb{R}_+)$, which in presence of (iii) entails

$$\|s\|_{\dot{H}_2^{\frac{1}{2}-\sigma}(\mathbb{R}_+)} \leq c\|s\|_{L_1(\mathbb{R}_+)}^{\sigma}\|s\|_{H_1^1(\mathbb{R}_+)}^{1-\sigma} \leq c\mu^{-\frac{\sigma}{p}}, \quad \mu \geq 1.$$

Here the last inequality is verified by (ii) and (iii), because

$$\|s\|_{H^1_1(\mathbb{R}_+)} = \|s\|_{L_1(\mathbb{R}_+)} + \|\dot{s}\|_{L_1(\mathbb{R}_+)} \le c(\mu^{-\frac{1}{\rho}} + 1) \le 2c,$$

provided that $\mu \ge 1$. Note that the latter implies, that $s \in \dot{H}_2^{\frac{1}{2}-\sigma}(\mathbb{R}_+)$ for all $\mu > 0$.

Let now $\theta \in [0, \sigma)$. Then $\sigma - \theta \in (0, \sigma] \subset (0,1)$ and the desired result follows by repeating the proof of assertion (iv) with replacing σ by $\sigma - \theta$. $\square$

Lemma 3.9. *Let $\mu > 0$, $\alpha \in (0, 2)$, $\beta > 0$, $\sigma \in (0, 1)$, $\theta \in [0, 1]$ and denote by $r : \mathbb{R}_+ \to \mathbb{R}$ the solution of the problem*

$$r(t) + \mu (g_\alpha * r)(t) = g_\beta(t), \qquad t \ge 0, \tag{3.6}$$

where g_κ means the Riemann-Liouville kernel of fractional integration; see (1.4). Then there is a constant $c > 0$ so that

$$\|r\|_{\dot{H}_2^{\theta+\frac{1}{2}-\sigma}(\mathbb{R}_+)} \le c\mu^{\frac{2(1-\beta+\theta-\sigma)}{\alpha}},$$

whenever $\beta \in (1-\sigma+\theta, 1-\sigma+\theta+\alpha)$.

Proof. By the Paley-Wiener theorem (cf. Theorem A.2), a function f belongs to $L_2(\mathbb{R})$ if and only if $\hat{f} \in \mathscr{H}_2(\mathbb{C}_+)$, the Hardy space of exponent 2, and the theorem also yields

$$\|f\|_{L_2(\mathbb{R})}^2 = \frac{1}{2\pi}\|\hat{f}\|_{\mathscr{H}(\mathbb{C}_+)}^2 = \frac{1}{2\pi}\int_\mathbb{R} |\hat{f}(i\rho)|^2 \, d\rho = \frac{1}{2\pi}\int_\mathbb{R} |(\mathcal{F}f)(\rho)|^2 \, d\rho.$$

Extending the function r trivially by zero for $t < 0$, we have by means of identity (1.7)

$$\|r\|_{\dot{H}_2^{\theta+\frac{1}{2}-\sigma}(\mathbb{R}_+)}^2 = \|\partial^{\theta+\frac{1}{2}-\sigma} r\|_{L_2(\mathbb{R})}^2 = \int_\mathbb{R} |\mathcal{F}r(\rho)|^2 \, |\rho|^{2\theta+1-2\sigma} d\rho = \int_\mathbb{R} |\hat{r}(i\rho)|^2 \, |\rho|^{2\theta+1-2\sigma} d\rho.$$

Observe now, that due to (3.6) we obtain a representation of r in terms of its Laplace transform, that is

$$\hat{r}(\lambda) = \frac{\hat{g}_\beta(\lambda)}{1+\mu\hat{g}_\alpha(\lambda)} = \frac{\lambda^\alpha}{\lambda^\beta(\lambda^\alpha+\mu)}, \qquad \operatorname{Re}\lambda \ge 0, \quad \lambda \ne 0.$$

3.2. PROOF OF THE MAIN RESULTS

Hence, we can proceed with

$$\|r\|^2_{\dot{H}_2^{\theta+\frac{1}{2}-\sigma}(\mathbb{R}_+)} = \int_{\mathbb{R}} |\widehat{r}(i\rho)|^2 |\rho|^{2\theta+1-2\sigma} d\rho$$

$$= \int_{\mathbb{R}} \left[\frac{|\rho|^\alpha}{|\rho|^\beta(|\rho|^\alpha+\mu)}\right]^2 |\rho|^{2\theta+1-2\sigma} d\rho$$

$$= 2\int_0^\infty \left[\frac{\rho^\alpha}{\rho^\beta(\rho^\alpha+\mu)}\right]^2 \rho^{2\theta+1-2\sigma} d\rho$$

$$= 2\mu^{\frac{2(1-\beta-\sigma)}{\alpha}} \int_0^\infty \left[\frac{\tau^{\theta+\alpha-\beta-\sigma+\frac{1}{2}}}{1+\tau^\alpha}\right]^2 d\tau$$

and the last integral is finite if and only if $1-\sigma+\theta < \beta < 1-\sigma+\theta+\alpha$. $\square$

3.2.1 Proof of Theorem 3.1

It is due to [88, Section I.1], that if (A) and (b) are valid, problem (3.1) admits a resolvent $S(t)$, so that $S \in L_1(\mathbb{R}_+; \mathcal{B}(\mathcal{H}))$, $S(t)$ is strongly continuous, is uniformly bounded by 1 and $\lim_{t \to \infty} |S(t)|_{\mathcal{B}(\mathcal{H})} = 0$. Consequently, the unique mild solution u of problem (3.1) exists and is defined[1] by

$$u(t) = \int_0^t S(t-\tau) d(Q^{1/2}\mathcal{X})(\tau), \quad t \geq 0. \qquad (3.7)$$

By means of the spectral decomposition of the operator A, the resolvent family decomposes into

$$S(t)x = \sum_{k=1}^\infty s_k(t)(x|a_n)a_n, \quad t \geq 0, \quad x \in \mathcal{H}, \qquad (3.8)$$

where the scalar functions $s_n(t) := s(t, \mu_n)$ are the solutions of the scalar problems

$$s_n(t) + \mu_n \int_0^t b(t-\tau)s_n(\tau)d\tau = 1, \quad t \geq 0. \qquad (3.9)$$

Observe now, that due to Hypothesis (A) there is a positive integer $N_\mu \in \mathbb{N}$, so that $\mu_n \geq 1$ for all $n > N_\mu$. Then Corollary 2.34 yields the existence of a constant $c > 0$ (in

---

[1] The notion of a mild solution is well-known for problems of the form (3.1) (e.g. [25]) and can be motivated from the variation of parameters formula (1.12).

the sequel generic) so that for $t \geq 0$ in view of Lemma 3.8 (iv) it is

$$
\begin{aligned}
\mathbb{E}\, |u(t)|_\mathcal{H}^2 &\leq c \sum_{k=1}^\infty |Q^{1/2}a_k|_\mathcal{H}^2 \|s_k\|_{\dot{\mathrm{H}}_2^{\frac{2-\gamma}{2}}([0,t])}^2 \\
&\leq c \sum_{k=1}^\infty |Q^{1/2}a_k|_\mathcal{H}^2 \|s_k\|_{\dot{\mathrm{H}}_2^{\frac{2-\gamma}{2}}(\mathrm{R}_+)}^2 \\
&\leq c \left[\sum_{k=1}^{N_\mu} |Q^{1/2}a_k|_\mathcal{H}^2 \|s_k\|_{\dot{\mathrm{H}}_2^{\frac{2-\gamma}{2}}(\mathrm{R}_+)}^2 + \sum_{k=N_\mu+1}^\infty |Q^{1/2}a_k|_\mathcal{H}^2 \mu_k^{\frac{1-\gamma}{\rho}} \right] \\
&= c \left[C(N_\mu) + \sum_{k=N_\mu+1}^\infty (Qa_k|a_k)_\mathcal{H} \mu_k^{\frac{1-\gamma}{\rho}} \right] \\
&= c \left[C(N_\mu) + \sum_{k=N_\mu+1}^\infty (Qa_k|A^{\frac{1-\gamma}{\rho}}a_k)_\mathcal{H} \right] \\
&\leq c \left[C(N_\mu) + \left\| QA^{\frac{1-\gamma}{\rho}} \right\|_{\mathscr{L}_1(\mathcal{H})} \right]
\end{aligned}
\tag{3.10}
$$

holds true. Then for $s, t \geq 0$ and $0 < \theta < \frac{\gamma-1}{2}$ the estimate

$$
\begin{aligned}
\mathbb{E}\, |u(t) - u(s)|_\mathcal{H}^2 &\leq c \sum_{k=1}^\infty |Q^{1/2}a_k|_\mathcal{H}^2 \left[\|s_k\|_{\dot{\mathrm{H}}_2^{\frac{2-\gamma}{2}}(\mathrm{R}_+)} + \|s_k\|_{\dot{\mathrm{H}}_2^{\theta+\frac{2-\gamma}{2}}(\mathrm{R}_+)} \right]^2 |t-s|^{2\theta} \\
&\leq c \left[C(N_\mu) + \sum_{k=N_\mu+1}^\infty (Qa_k|a_k)_\mathcal{H} \mu_k^{\frac{2\theta-\gamma+1}{\rho}} \right] |t-s|^{2\theta} \\
&\leq c \left[C(N_\mu) + \left\| QA^{\frac{2\theta-\gamma+1}{\rho}} \right\|_{\mathscr{L}_1(\mathcal{H})} \right] |t-s|^{2\theta}
\end{aligned}
\tag{3.11}
$$

yields the mean-square continuity of u. Lastly, we may employ the Kahane-Khinchine inequality (cf. Theorem A.3) to obtain for all $2 < p < \infty$

$$
\mathbb{E}|u(t)-u(s)|_\mathcal{H}^p \leq c_p \left(\mathbb{E}|u(t)-u(s)|_\mathcal{H}^2\right)^{\frac{p}{2}} \leq c_p \left[C(N_\mu) + \left\| QA^{\frac{2\theta-\gamma+1}{\rho}} \right\|_{\mathscr{L}_1(\mathcal{H})} \right]^{\frac{p}{2}} |t-s|^{p\theta} \tag{3.12}
$$

and the Kolmogorov-Čentsov-Theorem (cf. Theorem A.4) yields the claimed Hölderianity for every $\Theta \in (0, \theta - \frac{1}{p})$ for all $p \in (2, \infty)$.

3.2.2 Proof of Theorem 3.5

For $\alpha \in (0,2)$ and $\beta > 0$ problem (3.5) admits a resolvent $R(t)$ which decomposes into

$$
R(t)x = \sum_{k=1}^\infty r_k(t)(x|a_n)a_n, \qquad t \geq 0, \qquad x \in \mathcal{H}, \tag{3.13}
$$

3.2. PROOF OF THE MAIN RESULTS

where the scalar fundamental solutions $r_n(t) := r(t, \mu_n)$, $n \in \mathbb{N}$, of (3.5) can be expressed in terms of its Laplace transform

$$\widehat{r}_n(\lambda) = \frac{\widehat{g}_\beta(\lambda)}{1 + \mu_n \widehat{g}_\alpha(\lambda)} = \frac{\lambda^\alpha}{\lambda^\beta(\lambda^\alpha + \mu_n)}, \quad \operatorname{Re}\lambda \geq 0, \quad \lambda \neq 0, \quad \mu_n > 0. \quad (3.14)$$

Thus the mild solution u of problem (3.5) exists and is of the form

$$u(t) = \int_0^t R(t-\tau) \mathrm{d}(Q^{1/2}\mathcal{X})(\tau), \quad t \geq 0. \quad (3.15)$$

Then Corollary 2.34 ensures the existence of a constant $c > 0$ (in the sequel generic), so that in view of Lemma 3.9 we deduce for $t \geq 0$

$$\begin{aligned}
\mathbb{E}|u(t)|_\mathcal{H}^2 &\leq c \sum_{k=1}^\infty |Q^{1/2}a_k|_\mathcal{H}^2 \|r_k\|_{\mathrm{H}_2^{\frac{2-\gamma}{2}}([0,t])}^2 \leq c \sum_{k=1}^\infty |Q^{1/2}a_k|_\mathcal{H}^2 \|r_k\|_{\mathrm{H}_2^{\frac{2-\gamma}{2}}(\mathbb{R}_+)}^2 \\
&\leq c \sum_{k=1}^\infty |Q^{1/2}a_k|_\mathcal{H}^2 \mu_k^{\frac{3-2\beta-\gamma}{\alpha}} = c \sum_{k=1}^\infty (Qa_k|a_k)_\mathcal{H} \mu_k^{\frac{3-2\beta-\gamma}{\alpha}} \\
&= c \sum_{k=1}^\infty (Qa_k | A^{\frac{3-2\beta-\gamma}{\alpha}} a_k)_\mathcal{H} = c \left\| QA^{\frac{3-2\beta-\gamma}{\alpha}} \right\|_{\mathscr{L}_1(\mathcal{H})}.
\end{aligned} \quad (3.16)$$

In the same manner

$$\begin{aligned}
\mathbb{E}|u(t) - u(s)|_\mathcal{H}^2 &\leq c \sum_{k=1}^\infty |Q^{1/2}a_k|_\mathcal{H}^2 \left[\|r_k\|_{\mathrm{H}_2^{\frac{2-\gamma}{2}}(\mathbb{R}_+)} + \|r_k\|_{\mathrm{H}_2^{\theta+\frac{2-\gamma}{2}}(\mathbb{R}_+)}\right]^2 |t-s|^{2\theta} \\
&\leq c \sum_{k=1}^\infty (Qa_k|a_k)_\mathcal{H} \mu_k^{\frac{3-2\beta+2\theta-\gamma}{\alpha}} |t-s|^{2\theta} \\
&\leq c \left\| QA^{\frac{3-2\beta+2\theta-\gamma}{\alpha}} \right\|_{\mathscr{L}_1(\mathcal{H})} |t-s|^{2\theta}
\end{aligned} \quad (3.17)$$

holds for $s, t \in \mathbb{R}_+$ and $\theta \in [0, 1]$ and yields the mean-square continuity of u.

Again we may employ the Kahane-Khinchine inequality (cf. Theorem A.3) to obtain for all $2 < p < \infty$

$$\mathbb{E}|u(t) - u(s)|_\mathcal{H}^p \leq c_p \left(\mathbb{E}|u(t) - u(s)|_\mathcal{H}^2\right)^{\frac{p}{2}} \leq c_p \left\| QA^{\frac{3-2\beta+2\theta-\gamma}{\alpha}} \right\|_{\mathscr{L}_1(\mathcal{H})}^{\frac{p}{2}} |t-s|^{p\theta} \quad (3.18)$$

and the Kolmogorov-Čentsov-Theorem (cp. Theorem A.4) yields the claimed Hölderianity for every $\Theta \in (0, \theta - \frac{1}{p})$ for all $p \in (2, \infty)$. This completes the proof.

3.3 The case $\alpha = 2$

We conclude with a brief discussion of the case $\alpha = 2$. Then

$$\widehat{r}_n(\lambda) = \frac{\lambda^{2-\beta}}{\lambda^2 + \mu_n}, \quad n \in \mathbb{N},$$

(the functions r_n where introduced in (3.13)), hence there are poles $\pm i\sqrt{\mu_n}$ on the imaginary axis, and so Lemma 3.9 is not valid in this case. Therefore we proceed differently. It is shown in [25], that if $\frac{1}{2} < \beta < 3$ one obtains with the aid of the complex inversion formula for the Laplace transform

$$r_n(t) = \mu_n^{\frac{1-\beta}{2}} \left[\sin\left(\sqrt{\mu_n}t + \frac{(2-\beta)\pi}{2}\right) - \frac{1}{\pi}\sin((2-\beta)\pi) \int_0^\infty e^{-\sqrt{\mu_n}t\tau} \frac{\tau^{2-\beta}d\tau}{1+\tau^2} \right],$$

where $t > 0$. This formula shows in particular, that for every $t \in (0, T)$ it is $|r_n(t)| \leq c_T \mu_n^{\frac{1-\beta}{2}}$, where the constant c_T may depend on T (in the sequel generic). In case $\gamma > 2$ the embedding $L_{2/(\gamma-1)} \hookrightarrow \dot{H}_2^{\frac{2-\gamma}{2}}$ holds true and we obtain

$$|r_n|_{\dot{H}_2^\phi(0,T)} \leq c|r_n|_{\dot{H}_2^{\frac{2-\gamma}{2}}(0,T)} \leq c_T |r_n|_{L_{2/(\gamma-1)}(0,T)} \leq c_T \mu_n^{\frac{1-\beta}{2}},$$

for any fixed $T > 0$ and for all $n \in \mathbb{N}$. The condition for local existence in the case $\alpha = 2$ and $\gamma > 2$ is now immediate and reads as

$$\sum_{n=1}^\infty \nu_n \mu_n^{\frac{1-\beta}{2}} < \infty \quad \Longleftrightarrow \quad QA^{\frac{1-\beta}{2}} \in \mathscr{L}_1(\mathcal{H}).$$

Note, that this is not the limiting case of Theorem 3.5(i) as $\alpha \to 2$.

Chapter 4

Anomalous diffusion

Let $\alpha \in (0, 2)$, $G \subset \mathbb{R}^N$ to be a domain with boundary ∂G and $J = [0, T]$ a bounded time interval. We study the parabolic boundary problem of subdiffusion (if $\alpha < 1$), normal diffusion (if $\alpha = 1$), and superdiffusion (if $\alpha > 1$) with fractional stochastic disturbances on the boundary. This problem reads as

$$\begin{cases} \partial_t^\alpha u(t,x) - \Delta u(t,x) = 0, & t \in J, \quad x \in G, \\ \mathcal{D}u(t,x) = \psi(t,x), & t \in J, \quad x \in \partial G, \\ u(0,x) = 0, & x \in G, \end{cases} \quad (4.1)$$

in the basic space

$$V = L_2(J \times G \times \Omega),$$

where the boundary disturbance is modeled as

$$\psi(t, x, \omega) = \sum_{k=1}^{\infty} b_k(t, x) X_k(t, \omega) \quad (4.2)$$

and is suppose to satisfy

Hypothesis (ψ). $(X_i)_{i \in \mathbb{N}}$ are entirely independent processes subject to Hypothesis (ϕ) (see page 37) with a unique spectral density ϕ and $1 < \gamma < 3$. The multiplier $b := (b_n)_{n \in \mathbb{N}}$ is deterministic.

From time to time we will strengthen our assumptions and assume

Hypothesis (ψ_0). $(X_i)_{i\in\mathbb{N}}$ are entirely independent processes with a unique spectral density ϕ subject to Hypotheses (ϕ) (see page 37), so that $|\lambda|^\gamma \phi(\lambda) \equiv \text{const}$. Moreover, the multiplier $b := (b_n)_{n\in\mathbb{N}}$ is deterministic and for all $s, t \in J$ and $x \in \partial G$ we have $(b(t,x)|b(s,x))_{\ell_2} \geq 0$.

Remark 4.1. As shown in Section 2.6, centered Lévy processes and fractional Brownian motions satisfy Hypothesis (ψ_0). A fractional Riesz-Bessel motion $\mathcal{RB}^\beta_\eta$ is to Hypothesis (ψ_0) only if $\eta = 0$.

In order to circumvent trivial complications, we redefine the fractional derivative operator ∂^α as
$$(\partial^\alpha \phi)(t) := \frac{\mathrm{d}^2}{(\mathrm{d}t)^2} \int_0^t g_{2-\alpha}(t-\tau)\phi(\tau)\mathrm{d}\tau, \quad t \in \mathbb{R}_+.$$

With
$$Y = L_2(J \times \partial G \times \Omega)$$
we denote the basic space for the boundary process ψ. As to the operator $\mathcal{D}$, we either choose $\mathcal{D} = \partial_\nu$ which links to the Neumann problem, or $\mathcal{D} = I$ to study the Dirichlet problem. As usual I denotes the identity mapping.

Such problems arise in the theory of normal and anomalous diffusion, where the boundary conditions prescribe a stochastic inflow in case $\mathcal{D} = \partial_\nu$, and a stochastic concentration on the boundary for $\mathcal{D} = I$, respectively. Another typical application of those problems is the heat conduction in materials with memory (e.g. polymeric fluids or solids).

We are seeking for conditions on the parameter γ (determined by Hypothesis (ψ)) and properties of the pointwise multiplier b, so that the solution (see Section 4.2.1 for the present concept of a solution) u of (4.1) affiliates to the space Z_δ defined by

$$Z_\delta := {_0W_2^{\frac{\alpha\delta}{4}}}(J; L_2(G; L_2(\Omega))) \cap L_2\left(J; W_2^{\min\{\frac{\delta}{2};2\}}(G; L_2(\Omega))\right), \quad \delta \geq 0. \qquad (4.3)$$

It will turn out, that these spaces are appropriate solution spaces. Note that the class Z_4 appears as the maximal regularity class of type L_2 associated to problem (4.1). Moreover, the spaces Z_δ, with $\delta \geq 4$ are tailored to capture results with a higher time regularity. Higher spacial regularity is not treated in this thesis, since the resulting inevitable, purely technical, compatibility conditions cannot be motivated from the view of applications. For brevity we introduce the classes $U_{\delta,\gamma}$ and $U_{\delta,\gamma}^0$ for the pointwise multiplier $b := (b_i)_{i\in\mathbb{N}}$ as

$$U_{\delta,\gamma} := {}_0W_{2,\frac{\gamma-1}{2}}^{\frac{\alpha\delta}{4}}(J; L_2(\partial G; \ell_2)) \cap L_{2,\frac{\gamma-1}{2}}\left(J; W_2^{\frac{\delta}{2}}(\partial G; \ell_2)\right), \quad \delta \geq 0,$$
$$U_{\delta,\gamma}^0 := {}_0W_2^{\frac{\alpha\delta}{4}}(J; L_2(\partial G; \ell_2)) \cap L_{2,\frac{\gamma-1}{2}}\left(J; W_2^{\frac{\delta}{2}}(\partial G; \ell_2)\right), \quad \delta \geq 0. \quad (4.4)$$

It is then readily seen, that $U_{\delta,\gamma}^0 \hookrightarrow U_{\delta,\gamma}$.

4.1 Main results

In what follows let $\alpha \in (0,2)$ and $G \subset \mathbb{R}^N$ be either the N dimensional half-space, given by

$$\mathbb{R}_+^N := \{x := (x', y) \in \mathbb{R}^N : x' \in \mathbb{R}^{N-1}, y > 0\}, \quad (4.5)$$

or a domain with compact boundary ∂G of class $C^{\lceil 2/\alpha\rceil+1}$, if not indicated otherwise.

Theorem 4.2. *Assume Hypothesis (ψ) holds. Let $0 \leq \nu < \frac{2(\gamma-1)}{\alpha}$ and in case $G \neq \mathbb{R}_+^N$ let $\nu \in [0, \frac{2(\gamma-1)}{\alpha}) \cap [0,4)$. Then the following hold if $b \in U_{\nu,\gamma}^0$, given by (4.4).*

(i) *The Dirichlet problem (4.1), i.e. $\mathcal{D} = I$, admits a unique solution u in the regularity class $Z_{\nu+1}$ given by (4.3). If, in addition, $\nu \leq 3$ and Hypothesis (ψ_0) is valid, then membership of b to the class $U_{\nu,\gamma}$ is necessary and sufficient.*

(ii) *The Neumann problem (4.1), i.e. $\mathcal{D} = \partial_\nu$, admits a unique solution u in the regularity class $Z_{\nu+3}$ given by (4.3). If, in addition, $\nu \leq 1$ and Hypothesis (ψ_0) is valid, then membership of b to the class $U_{\nu,\gamma}$ is necessary and sufficient.*

Note that the additional assumption $\nu < 4$, when $G \neq \mathbb{R}^N_+$, is only restrictive in the case of subdiffusion, if $\alpha < \frac{\gamma-1}{2}$. However, in view of maximal L_2-regularity it is not obstructive at all. If one seeks for strong solutions of problem (4.1) with Dirichlet boundary condition, Theorem 4.2 shows that one necessarily has to assume that $\gamma > \frac{3\alpha+2}{2}$. On the other hand, in view of the Neumann problem the above theorem yields the existence of strong solutions, only if $\gamma > \frac{\alpha+2}{2}$. The following corollaries concern a result on mixed regularity classes with either full spatial regularity or full temporal regularity. In the Dirichlet case, i.e. $\mathcal{D} = I$, this results reads as

Corollary 4.3. The Dirichlet problem (4.1) admits a unique solution u and

(i) $u \in {}_0W_2^{\vartheta\alpha}(J; W_2^2(G; L_2(\Omega)))$, $\vartheta \geq 0$,

(ii) $u \in {}_0W_2^{\alpha}(J; W_2^{2\vartheta}(G; L_2(\Omega)))$, $\vartheta \in [0,1]$,

provided that

(a) $\vartheta < \frac{\gamma-1}{2\alpha} - \frac{3}{4}$;

(b) $\vartheta < \frac{1}{4}$, in case $G \neq \mathbb{R}^N_+$;

(c) $b \in U^0_{4\vartheta+3,\gamma}$.

Proof. Set $\nu = 4\vartheta + 3$, then clearly $3 \leq \nu < \frac{2(\gamma-1)}{\alpha}$ and, in addition, $\nu < 4$ if $G \neq \mathbb{R}^N_+$, thus Theorem 4.2 yields $u \in Z_{\nu+1}$. By the mixed derivative theorem we obtain

$$Z_{\nu+1} \hookrightarrow {}_0W_2^{\frac{\alpha(\nu+1)}{4}\theta}\left(J; W_2^{\frac{\nu+1}{2}(1-\theta)}(G; L_2(\Omega))\right), \quad \theta \in [0,1]$$

and the choice $\theta = \frac{\nu-3}{\nu+1}$ proves assertion (i), while $\theta = \frac{4}{\nu+1}$ gives (ii). □

In case of a boundary condition of Neumann type, i.e. $\mathcal{D} = \partial_\nu$, we deduce

4.1. MAIN RESULTS

Corollary 4.4. The Neumann problem (4.1) admits a unique solution u and

(i) $u \in {}_0W_2^{\vartheta\alpha}(J; W_2^2(G; L_2(\Omega)))$, $\quad \vartheta \geq 0$,

(ii) $u \in {}_0W_2^{\alpha}(J; W_2^{2\vartheta}(G; L_2(\Omega)))$, $\quad \vartheta \in [0, 1]$,

provided that

(a) $\vartheta < \frac{\gamma-1}{2\alpha} - \frac{1}{4}$;

(b) $\vartheta < \frac{3}{4}$, in case $G \neq \mathbb{R}_+^N$;

(c) $b \in U_{4\vartheta+1,\gamma}^0$.

Proof. Repeat the arguments of the proof of Corollary 4.3 with $\nu = 4\vartheta + 1$. □

Here we discuss mainly the $L_2(\Omega)$-valued case. The subsequent proposition – which is easy to prove but never the less useful – covers, in the presence of Theorem 4.2, a result in the pathwise sense.

Proposition 4.5. Let u belong to the space Z_δ given by (4.3) and $s \geq 0$ a real number.

(i) If $\delta > \frac{2}{\alpha}(2s+1)$, then $u \in L_2(\Omega; C^s(J; L_2(G)))$.

(ii) If $2s + N < \delta \leq 4$, then $u \in L_2(\Omega; L_2(J; C^s(G)))$.

(iii) If $\delta > 4$ and $s < \frac{4-N}{2}$, then $u \in L_2(\Omega; L_2(J; C^s(G)))$.

(iv) If $\delta > \max\left\{4; \frac{8(2s+N+1)}{\alpha(3-2s-N)}\right\}$ and $s < \frac{3-N}{2}$, then $u \in L_2(\Omega; C^s(J \times G))$.

Proof. Fubini's Theorem yields

$$Z_\delta = L_2(\Omega; {}_0W_2^{\frac{\alpha\delta}{4}}(J; L_2(G))) \cap L_2(\Omega; L_2(J; W_2^{\min\{\frac{\delta}{2};2\}}(G)))$$

and it is due to Sobolev imbedding that $_0W_2^\theta(K) \hookrightarrow C^s(K)$ if $s < \theta - \frac{\dim K}{2}$. Then a simple computation confirms (i), (ii) and (iii). Turning to (iv) we allude to

$$Z_\delta \hookrightarrow L_2\left(\Omega; W_2^{\frac{2\alpha\delta}{\alpha\delta+8}}(J \times G)\right) \quad \text{if} \quad \delta > 4,$$

which is due to the mixed derivative theorem and the claim follows via Sobolev imbedding. □

It is worthwhile to mention that in view of Proposition 4.5, there exists a feasible δ for (ii) and (iii) only if $N \leq 3$, and for (iv) only if $N \leq 2$. The remaining part of this chapter is devoted to the proof of Theorem 4.2. It is organized as follows. Next, in Section 4.2.1, we provide the notion of a weak solution associated to problem (4.1). Then, in Section 4.2.2, we prove Theorem 4.2 in the half-space setting. By means of spatial localization, which is the aim of Section 4.2.3, we will carry over the half-space results to domains in Section 4.2.4.

4.2 Proof of the main results

4.2.1 Weak solutions

By g_κ, we denote the Riemann-Liouville kernel; see (1.4). We call a function $u \in X$ weak solution of the Dirichlet problem (4.1), i.e. $\mathcal{D} = I$, if it satisfies the integral equation

$$-\int_G \int_J (\partial_t^2 \phi_D)(g_{2-\alpha} * u)\mathrm{d}t\mathrm{d}x + \int_G \int_J (\Delta \phi_D) u \mathrm{d}t\mathrm{d}x = \int_{\partial G} \int_J (\partial_\nu \phi_D)\psi \mathrm{d}t\mathrm{d}x \quad (4.6)$$

for all test functions ϕ_D in the class

$$\{\phi_D \in W_2^2(G; L_2(J)) : \phi_D|_{\partial G} = 0\} \cap \{\phi_D \in W_2^2(J; L_2(G)) : \phi_D(T) = \partial_t \phi_D(T) = 0\}. \quad (4.7)$$

Similarly, we call a function $u \in V$ weak solution of the Neumann problem (4.1), i.e. $\mathcal{D} = \partial_\nu$, if is satisfies the integral equation

$$-\int_G \int_J (\partial_t^2 \phi_N)(g_{2-\alpha} * u)\mathrm{d}t\mathrm{d}x + \int_G \int_J (\Delta \phi_N) u \mathrm{d}t\mathrm{d}x = \int_{\partial G} \int_J \phi_N \psi \mathrm{d}t\mathrm{d}x \quad (4.8)$$

4.2. PROOF OF THE MAIN RESULTS

for all test functions ϕ_N in the class

$$\{\phi_N \in W_2^2(G; L_2(J)) : \partial_\nu \phi_N \mid_{\partial G} = 0\} \cap \{\phi_N \in W_2^2(J; L_2(G)) : \phi_N(T) = \partial_t \phi_N(T) = 0\}. \tag{4.9}$$

Equations (4.6) resp. (4.8) can be obtained by multiplying problem (4.1) with ϕ_D resp. ϕ_N and integrating over J and G. Note that by construction every strong solution is also a weak solution. The converse is not true in general. Observe that the classes (4.7) and (4.9) are nontrivial and dense in V, since they contain the C^∞-functions with compact support in $(0, T) \times G$.

In the half-space setting, that is if $G = \mathbb{R}_+^N$, one achieves a more explicit representation of a weak solution of problem (4.1). To this purpose we define the operator

$$F := (\partial_t^\alpha - \Delta_{x'})^{1/2} \tag{4.10}$$

acting on the basic space Y with domain

$$D(F) = {}_0W_2^{\frac{\alpha}{2}}(J; L_2(\mathbb{R}^{N-1}; L_2(\Omega))) \cap L_2(J; W_2^1(\mathbb{R}^{N-1}; L_2(\Omega))). \tag{4.11}$$

Since the operator ∂_t^α is sectorial of angle $\frac{\alpha\pi}{2}$ and, moreover, commutes with the negative Laplacian $-\Delta_{x'}$ it is due to the Kalton-Weis-Theorem [59, Theorem 6.3], that the operator F is sectorial of angle $\frac{\alpha\pi}{4}$, hence is the negative generator of an analytic C_0-semigroup, provided $0 < \alpha < 2$.

Let $\Lambda : \{\partial_\nu, I\} \to \{0, 1\}$ be the function which indicates the Neumann problem; precisely $\Lambda_\mathcal{D} := \Lambda(\mathcal{D}) = 1$ if and only if $\mathcal{D} = \partial_\nu$. We are now in the position to rewrite problem (4.1) in coordinates according to (4.5) as the ordinary differential equation

$$\begin{cases} -\partial_y^2 u(y) + F^2 u(y) = 0, & y > 0, \\ (1 - \Lambda_\mathcal{D})u(0) - \Lambda_\mathcal{D} \partial_y u(0) = \psi. \end{cases} \tag{4.12}$$

The deterministic case (cf. [89, Section 3]) gives raise to call a function u a (weak) solution of (4.12), if it satisfies

$$u(y) = e^{-Fy} F^{-\Lambda_\mathcal{D}} \psi, \quad t > 0, \tag{4.13}$$

where as usual $F^0 := I$. Here e^{tA} denotes the analytic C_0-semigroup generated by the operator A. In particular, this formula depicts the well-posedness of problem (4.1) in the sense of Hadamard, i.e. the problem admits a unique solution which depends continuously on the data, in some reasonable topology.

In order to show, that a weak solution of the form (4.13) satisfies the representation formula (4.6) resp. (4.8), we make use of an approximation argument. We exemplify this argument for the case of a boundary condition of Dirichlet type. To this end let ψ_n belong to

$$D(F^{\frac{3}{2}}) = {}_0W_2^{\frac{3\alpha}{4}}(J; L_2(\mathbb{R}^{N-1}; L_2(\Omega))) \cap L_2(J; W_2^{\frac{3}{2}}(\mathbb{R}^{N-1}; L_2(\Omega)))$$

for all $n \in \mathbb{N}$ so that $\psi_n \to \psi \in Y$ as n tends to infinity. Theorem 4.2 yields that the function $u_n(y) = e^{-Fy}\psi_n$ affiliates to the class Z_4, hence is a strong solution of the Dirichlet problem

$$\begin{cases} \partial_t^\alpha u_n(t,x) - \Delta u_n(t,x) = 0, & t \in J, \ x \in \mathbb{R}_+^N, \\ u_n(t,x) = \psi_n(t,x), & t \in J, \ x \in \mathbb{R}^{N-1}, \\ u_n(0,x) = 0, & x \in \mathbb{R}_+^N \end{cases}$$

for every $n \in \mathbb{N}$. It is due to the C_0-property of the semigroup e^{-Fy} and Theorem 4.2 that $u_n \to u \in Z_1$ as $n \to \infty$ and in particular by maximal regularity (the functions u_n are strong solutions for all $n \in \mathbb{N}$) and representation (4.6) we have the validity of the integral equation

$$-\int_{\mathbb{R}_+^N}\int_J (\partial_t^2\phi_D)(g_{2-\alpha} * u_n)\mathrm{d}t\mathrm{d}x + \int_{\mathbb{R}_+^N}\int_J (\Delta\phi_D)u_n\mathrm{d}t\mathrm{d}x = \int_{\mathbb{R}^{N-1}}\int_J (\partial_\nu\phi_D)\psi_n\mathrm{d}t\mathrm{d}x$$

for all $n \in \mathbb{N}$. Passing n to the limit we see that in the half-space setting, a weak solution of the form (4.13) satisfies equation (4.6). In this sense formulae (4.13) and (4.6) resp. (4.8) are connected.

We are now in the position to proof the main result in the half-space setting.

4.2.2 Proof of Theorem 4.2: Half-space setting.

Let $G = \mathbb{R}_+^N$, given by (4.5). The unique existence of a solution u of problem (4.1) is clear by (4.13). By Theorems 2.22 and 2.25 we have the implication

$$b \in U_{\nu,\gamma}^0 \implies \psi \in D(F^{\frac{\nu}{2}}),$$

4.2. PROOF OF THE MAIN RESULTS

where the operator F is given by (4.10) and

$$D(F^\theta) = {}_0W_2^{\frac{\alpha\theta}{2}}(J; L_2(\mathbb{R}^{N-1}; L_2(\Omega))) \cap L_2(J; W_2^\theta(\mathbb{R}^{N-1}; L_2(\Omega))), \qquad (4.14)$$

for $\theta \geq 0$. If, in addition, Hypothesis (ψ_0) is valid, then Theorems 2.22 and 2.25 enforce

$$b \in U_{\nu,\gamma} \iff \psi \in D(F^{\frac{\nu}{2}}).$$

Assertion (i) is proven, if we can show that $\psi \in D(F^{\frac{\nu}{2}})$ is equivalent to $u \in Z_{\nu+1}$. To this end we denote by $z \in \widetilde{V} := L_2(\mathbb{R} \times \mathbb{R}_+^N \times \Omega)$ the solution of the problem

$$\begin{cases} -\partial_y^2 z(y) + \widetilde{F}^2 z(y) = 0, & y > 0, \\ z(0) = \Psi, \end{cases}$$

where the process Ψ belongs to $\widetilde{Y} := L_2(\mathbb{R} \times \mathbb{R}^{N-1} \times \Omega)$ so that $\Psi\big|_{t \in J} = \psi$ holds and we define

$$\widetilde{F} := \sqrt{\partial_t^\alpha - \Delta_{x'} + I}$$

with domain

$$D(\widetilde{F}) = {}_0W_2^{\frac{\alpha}{2}}(\mathbb{R}; L_2(\mathbb{R}^{N-1}; L_2(\Omega))) \cap L_2(\mathbb{R}; W_2^1(\mathbb{R}^{N-1}; L_2(\Omega))).$$

Recall that by (4.13) z is of the form $z(y) := e^{-\widetilde{F}y}\Psi$ with $y \geq 0$.

In what follows $\mathcal{F}$ means the Fourier transform with respect to time t and tangential variable x'. Let $m = m(\lambda, \xi) = \sqrt{\lambda^\alpha + |\xi|^2 + 1}$ with $\lambda = i\rho$, $\rho \in \mathbb{R}$, $\xi \in \mathbb{R}^{N-1}$, denote the Fourier symbol of $\widetilde{F}(t, x')$. Suppressing the argument $\omega \in \Omega$, Plancherel's Theorem (cf. Theorem A.1) yields

$$\|\widetilde{F}^{\frac{\nu+1}{2}}z\|_{\widetilde{V}}^2 = \|\mathcal{F}\{\widetilde{F}^{\frac{\nu+1}{2}}z\}\|_{\widetilde{V}}^2 = \int_0^\infty \int_{\mathbb{R}} \int_{\mathbb{R}^{N-1}} |m^{\frac{1}{2}} \mathcal{F}\{\widetilde{F}^{\frac{\nu}{2}}z(y)\}(\lambda,\xi)|^2 d\xi d\rho dy$$

$$= \int_{\mathbb{R}} \int_{\mathbb{R}^{N-1}} \int_0^\infty |m| e^{-2\operatorname{Re} my} |\mathcal{F}\{\widetilde{F}^{\frac{\nu}{2}}\Psi\}(\lambda,\xi)|^2 dy d\xi d\rho$$

$$= \int_{\mathbb{R}} \int_{\mathbb{R}^{N-1}} \frac{|m|}{2\operatorname{Re} m} |\mathcal{F}\{\widetilde{F}^{\frac{\nu}{2}}\Psi\}(\lambda,\xi)|^2 d\xi d\rho.$$

Observe now, that due to $\alpha \in (0,2)$ the symbol m takes values in an open sector of the complex plane, symmetric with respect to the positive real half axis $\mathbb{R}_+$, with vertex 0 and opening angle $\vartheta < \pi$. This captures the existence of constants $c_1, c_2 > 0$, such that

$$c_1|m| \leq \operatorname{Re} m \leq c_2|m|$$

holds. Therefrom we obtain for $\nu \in [0, \frac{2(\gamma-1)}{\alpha}) \cap [0,3]$

$$\|\widetilde{F}^{\frac{\nu}{2}}\Psi\|_{\widetilde{Y}}^2 \leq c \int_{\mathbb{R}} \int_{\mathbb{R}^{N-1}} \frac{|m|}{2\operatorname{Re} m} |\mathcal{F}\{\widetilde{F}^{\frac{\nu}{2}}\Psi\}(\lambda,\xi)|^2 \mathrm{d}\xi \mathrm{d}\rho = \|\widetilde{F}^{\frac{\nu+1}{2}}z\|_V^2$$

which is the key to necessity. Turning to sufficiency we deduce for $\nu \in [0, \frac{2(\gamma-1)}{\alpha})$

$$\|\widetilde{F}^{\frac{\nu+1}{2}}z\|_V^2 \leq c \int_{\mathbb{R}} \int_{\mathbb{R}^{N-1}} |\mathcal{F}\{\widetilde{F}^{\frac{\nu}{2}}\Psi\}(\lambda,\xi)|^2 \mathrm{d}\xi \mathrm{d}\rho = c\|\widetilde{F}^{\frac{\nu}{2}}\Psi\|_{\widetilde{Y}}^2,$$

hence $\bar{z} := z\mid_{t\in J} \in Z_{\nu+1}$, where $\bar{z} = z\mid_{t\in J}$ denotes the restriction of z to J. Observe now, that $u = \bar{z} + w$, where $w \in V$ is the solution of the problem

$$\begin{cases} -\partial_y^2 w(y) + F^2 w(y) = \bar{z}, & y > 0, \\ w(0) = 0, \end{cases}$$

with F given by (4.10). It is due to [112, Theorem 3.1] that

$$w \in {}_0W_2^{\alpha + \frac{\alpha(\nu+1)}{4}}(J; L_2(G; L_2(\Omega))) \cap L_2\left(J; {}_0W_2^2(G; L_2(\Omega))\right) = Z_{\nu+5},$$

which in turn yields $u \in Z_{\nu+1}$ and assertion (i) is proven.

Turning to (ii) let us denote by v the solution of the Dirichlet problem (4.1). Recall that by (4.13) it is $u = F^{-1}v$, and so in particular we have

$$v \in Z_{\nu+1} \iff u \in Z_{\nu+3}$$

for all $0 \leq \nu < \frac{2(\gamma-1)}{\alpha}$. Employing (i) completes the proof.

4.2.3 Spatial localization

Let now $G \subset \mathbb{R}^N$ be a domain with compact boundary ∂G of class C^2. In case G is unbounded one has to think of an exterior domain. Since the problem under investigation is fully known in the space $\mathbb{R}^N$ (e.g. Zacher [113, Theorem 3.1]) and, by the

4.2. PROOF OF THE MAIN RESULTS

above, in the half-space $\mathbb{R}^N_+$ the apparent strategy is to localize problem (4.1) and to apply the known results.

In order to prevent the localization with respect to time we do consider the following two auxiliary problems

$$\begin{cases} \partial_t^\alpha v(t,x) + (\lambda - \Delta)v(t,x) = 0, & t \in J, \quad x \in G, \\ \mathcal{D}v(t,x) = \psi(t,x), & t \in J, \quad x \in \partial G, \\ v(0,x) = 0, & x \in G, \end{cases} \quad (4.15)$$

where $\lambda > 0$ is chosen later and

$$\begin{cases} \partial_t^\alpha w(t,x) - \Delta w(t,x) = \lambda v(t,x), & t \in J, \quad x \in G, \\ \mathcal{D}w(t,x) = 0, & t \in J, \quad x \in \partial G, \\ w(0,x) = 0, & x \in G. \end{cases} \quad (4.16)$$

Note, that the function $u = v + w$ clearly solves the initial problem (4.1). Our strategy is as follows. We are going to localize problem (4.15) with respect to space and obtain a solution on the whole of $[0,T]$ by choosing λ sufficiently large. Then, depending on the resulting regularity of v, the regularity of w is known by [113, Theorem 3.4].

Since the technique of localization is well-known (e.g. Denk et al. [34, Section 8]) we just go briefly through the prearrangements. Let $x_0 \in \partial G$ be an arbitrary element of the boundary. Without loss of generality, we may assume that $x_0 = 0$ and the outer normal at x_0 satisfies $n(x_0) = (0, \ldots, 0, -1)$. This can always be achieved by a composition of a translation and a rotation in $\mathbb{R}^N$. Such affine mappings of $\mathbb{R}^N$ onto itself clearly leave all function spaces under consideration invariant. By definition of a C^2-boundary there is an open neighborhood $U = U_1 \times U_2 \subset \mathbb{R}^N$ of x_0 with $U_1 \subset \mathbb{R}^{N-1}$ and $U_2 \subset \mathbb{R}$ as well as a function $\zeta \in C^2(U_1)$, such that

$$\partial G \cap U = \{x = (x', y) \in U : y = \zeta(x')\},$$
$$G \cap U = \{x = (x', y) \in U : y > \zeta(x')\}.$$

Using now the notation $x = (x_1, \ldots, x_N)$ we define $\vartheta : U \to \mathbb{R}^N$ in virtue of

$$\vartheta_k(x) = \begin{cases} x'_k & : k = 1, \ldots, N-1 \\ y - \zeta(x') & : k = N \end{cases}. \quad (4.17)$$

It is easy to see, that $\vartheta \in C^2(U; \mathbb{R}^N)$ is one-to-one and satisfies $G \cap U = \{x \in U : \vartheta_N(x) > 0\}$ as well as $\partial G \cap U = \{x \in U : \vartheta_N(x) = 0\}$. Observe, that the function ζ can be extended to a function in $C^2(\mathbb{R}^{N-1})$ with compact support. For brevity we denote the extension of ζ again by ζ.

Regarding spatial localization, by the boundedness of ∂G, there exists a radius $r_0 > 0$ such that ∂G is entirely contained in the open ball $B_{r_0}(0)$. If G is unbounded we set $U_0 = \{x \in \mathbb{R}^N; |x| > r_0\}$, otherwise we may assume that $\overline{G} \subset B_{r_0}(0)$ and put $U_0 = \emptyset$. Now, we cover $\overline{B_{r_0}(0)}$ by finitely many open sets U_j, $j = 1, \ldots, n$, which are subject to

(U1) $U_j \cap \partial G = \emptyset$ and $U_j = B_{r_j}(x_j)$ for all $j = 1, \ldots n_1$.

(U2) $U_j \cap \partial G \neq \emptyset$ for $j = n_1 + 1, \ldots, n$ and there exists $x_j \in U_j \cap \partial G$ and $\zeta_j \in C^2(\mathbb{R}^{N-1})$ with compact support such that $U_j \cap \partial G = \{x = (x', y) \in U_j : y = \zeta_j(x')\}$ as well as $U_j \cap G = \{x = (x', y) \in U_j : y > \zeta_j(x')\}$, and $U_j = \vartheta_j^{-1}(B_{r_j}(x_j))$.

In what follows we denote by $\{\varphi_j\}_{j=0}^n \subset C^\infty(\mathbb{R}^N; [0, 1])$ a partition of the unity such that $\sum_{j=0}^n \varphi_j(x) \equiv 1$ on $\overline{G}$ and $\operatorname{supp} \varphi_j \subset U_j$. Observe now, that v solves (4.15) if and only if

$$\begin{cases} \partial_t^\alpha(\varphi_j v) + (\lambda - \Delta)(\varphi_j v) = -[\Delta, \varphi_j]v, & \text{in } J \times G, \quad j = 0, \ldots, n \\ \mathcal{D}(\varphi_j v) = \varphi_j \psi + [\mathcal{D}, \varphi_j]v, & \text{on } J \times \partial G, \ j = n_1 + 1, \ldots, n, \\ \varphi_j v \mid_{t=0} = 0. \end{cases}$$

In case $j = 0, \ldots, n_1$ we have to consider full-space problems for the functions $\varphi_j v$, for which the existence of the corresponding solution operators $\mathcal{S}_j^F$ is known. One obtains

$$\varphi_j v = \mathcal{S}_j^F(-[\Delta, \varphi_j]v) =: h_j^F(v), \quad j = 0, \ldots, n_1. \tag{4.18}$$

For $j = n_1 + 1, \ldots, n$, we get problems on crooked half-spaces with inhomogeneous Neumann or Dirichlet boundary condition. Using the common affine mappings that, in particular, transform x_j to the origin combined with an appropriate variable trans-

4.2. PROOF OF THE MAIN RESULTS

formation and denoting by Γ_y the trace operator at $y = 0$ leads to

$$\begin{cases} \partial_t^\alpha \Theta_j^{-1}(\varphi_j v) + (\lambda - \Delta)^{\vartheta_j} \Theta_j^{-1}(\varphi_j v) = -\Theta_j^{-1}[\Delta, \varphi_j]v, & J \times \mathbb{R}_+^N, \\ \mathcal{D}^{\vartheta_j} \Theta_j^{-1}(\varphi_j v) = \Theta_j^{-1}(\varphi_j \psi) + \Theta_j^{-1}\Gamma_y[\mathcal{D}, \varphi_j]v, & J \times \mathbb{R}^{N-1}, \\ \Theta_j^{-1}(\varphi_j v)\mid_{t=0} = 0, \end{cases}$$

that is, to half-space problems for $\Theta_j^{-1}(\varphi_j v)$. Here the pull-back $\Theta_j v$ is defined on int G by $\Theta_j v(x) = v(\vartheta_j(x))$ and $(\lambda - \Delta)^{\vartheta_j} := \lambda - \Theta_j^{-1}\Delta\Theta_j$ as well as $\mathcal{D}^{\vartheta_j} := \Theta_j^{-1}\mathcal{D}\Theta_j$. Choosing the radii r_i, $i = 1, \ldots, n$, sufficiently small, Theorem 4.2 in connection with a perturbation argument asserts the existence of solution operators $\mathcal{S}_j^B$ for the above problems. So we immediately get

$$\varphi_j v = \Theta_j \mathcal{S}_j^B \begin{pmatrix} -\Theta_j^{-1}[\Delta, \varphi_j]v \\ \Theta_j^{-1}(\varphi_j \psi) + \Theta_j^{-1}\Gamma_y[\mathcal{D}, \varphi_j]v \end{pmatrix} =: h_j^B(\psi, v), \qquad (4.19)$$

for $j = n_1 + 1, \ldots, n$. Summing now over all j yields the formula

$$v = \sum_{j=0}^{n_1} h_j^F(v) + \sum_{j=n_1+1}^{n} h_j^B(\psi, v) =: \mathcal{G}(v) + \mathcal{K}(\psi), \qquad (4.20)$$

which is necessary for v to be a solution of (4.1). To see that (4.20) is also sufficient we refer to Zacher [111]. Summarizing, we deduced a fixed point equation (4.20) for v, where the first sum is determined by the data, and the second contains only terms of lower order. By means of the contraction principle, this fixed point equation can be solved on $J = [0, T]$ provided $\mathcal{G}$ is a strict contraction on J. But this can always be arranged by choosing λ sufficiently large.

Before focusing the concrete Neumann or Dirichlet case, we prove an extremely useful result.

Proposition 4.6. Let $X = L_2(J \times \mathbb{R}_+^N \times \Omega)$ and $F = \sqrt{\partial_t^\alpha - \Delta_{x'}}$ with domain $D(F)$ given by (4.11) in coordinates according to (4.5). Then there is a constant $c > 0$, such that for $\nu \geq 0$ and sufficiently large $\lambda \geq 0$ we have

$$\left\| (F^2 + \lambda)^{\frac{\nu+1}{4}} e^{-(F^2+\lambda)^{\frac{1}{2}}y} \right\|_{\mathcal{B}(D(F^{\frac{\nu}{2}});X)}^2 \leq c(1 + \lambda^{\frac{\nu}{2}}).$$

Proof. Let $g \in D(F^{\frac{\nu}{2}})$ with $\nu \geq 0$ and $Y = L_2(J \times \mathbb{R}^{N-1} \times \Omega)$. Then by means of Plancherel's Theorem (cf. Theorem A.1) and Fourier transform with respect to time and space it is

$$\left\|(F^2+\lambda)^{\frac{\nu+1}{4}}e^{-(F^2+\lambda)^{\frac{1}{2}}y}g\right\|_X^2$$
$$\leq \int_{\mathbb{R}_+}\int_{\mathbb{R}^{N-1}}\int_{\mathbb{R}} \left||(i\tau)^\alpha+|\xi|^2+\lambda\right|^{\frac{\nu+1}{2}}e^{-2\operatorname{Re}\sqrt{(i\tau)^\alpha+|\xi|^2+\lambda}\,y}|\tilde{g}(i\tau,\xi)|^2 d\tau d\xi dy$$
$$= \int_{\mathbb{R}}\int_{\mathbb{R}^{N-1}} \left(\int_{\mathbb{R}_+} e^{-2\operatorname{Re}\sqrt{(i\tau)^\alpha+|\xi|^2+\lambda}\,y}dy\right) \left||(i\tau)^\alpha+|\xi|^2+\lambda\right|^{\frac{\nu+1}{2}}|\tilde{g}(i\tau,\xi)|^2 d\xi d\tau.$$

Observe now that for $\alpha \in (0,2)$ and sufficiency large λ the symbol $\sqrt{(i\tau)^\alpha+|\xi|^2+\lambda}$ takes values in the open sector $\Sigma(0,\frac{\pi}{2})$, so that we have in particular $c|\sqrt{(i\tau)^\alpha+|\xi|^2+\lambda}| \leq \operatorname{Re}\sqrt{(i\tau)^\alpha+|\xi|^2+\lambda}$ with a constant $c > 0$ (in the sequel generic) being independent of λ. Thus

$$\left\|(F^2+\lambda)^{\frac{\nu+1}{4}}e^{-(F^2+\lambda)^{\frac{1}{2}}y}g\right\|_X^2 \leq c\int_{\mathbb{R}}\int_{\mathbb{R}^{N-1}} \left[\left||(i\tau)^\alpha+|\xi|^2+\lambda\right|^{\frac{\nu}{4}}|\tilde{g}(i\tau,\xi)|\right]^2 d\xi d\tau$$
$$\leq c\int_{\mathbb{R}}\int_{\mathbb{R}^{N-1}} \left[[|(i\tau)^\alpha+|\xi|^2|^{\frac{\nu}{4}}+\lambda^{\frac{\nu}{4}}]|\tilde{g}(i\tau,\xi)|\right]^2 d\xi d\tau$$
$$\leq c\|g\|_{D(F^{\frac{\nu}{2}})}^2 + \lambda^{\frac{\nu}{2}}\|g\|_Y^2 \leq c(1+\lambda^{\frac{\nu}{2}})\|g\|_{D(F^{\frac{\nu}{2}})}^2$$

follows and the proof is complete. □

4.2.4 Proof of Theorem 4.2: Setting for domains.

This time we are first facing assertion (ii). Let us shortly recall what we have done in the preview. We split up the initial problem (4.1) in the two auxiliary problems (4.15) and (4.16), so that it suffices to seek for the regularity of the solution v of the localized version of (4.15). This is what remains to do. To this purpose we denote by $\mathcal{Z}_\nu^\lambda$ the space Z_ν equipped with the norm

$$\|\cdot\|_{\mathcal{Z}_\nu^\lambda} := \|\cdot\|_{Z_\nu} + \lambda\|\cdot\|_V, \qquad (4.21)$$

with λ from (4.15), where the space Z_ν is given by (4.3) with an admissible ν. Thanks to our preview it remains to show, that $\mathcal{G}$ is a strict contraction on J for $v \in \mathcal{Z}_{\nu+3}^\lambda$.

4.2. PROOF OF THE MAIN RESULTS

Thus we proceed as follows. Let v_1 and v_2 belong to $\mathcal{Z}_{\nu+3}^\lambda$, then the linearity of the solution operators captures

$$\|\mathcal{G}(v_1) - \mathcal{G}(v_2)\|_{\mathcal{Z}_{\nu+3}^\lambda} \leq \sum_{j=0}^{n_1} \|S_j^F([\Delta,\varphi_j](v_2 - v_1))\|_{\mathcal{Z}_{\nu+3}^\lambda} + \\ + \sum_{j=n_1+1}^{n} \left\| S_j^B \begin{pmatrix} \Theta_j^{-1}[\Delta,\varphi_j](v_2 - v_1) \\ \Theta_j^{-1}\Gamma_y[\mathcal{D},\varphi_j](v_1 - v_2) \end{pmatrix} \right\|_{\mathcal{Z}_{\nu+3}^\lambda}. \quad (4.22)$$

Turning to the first sum we observe, that for every $\varepsilon_1 > 0$ there is a constant c_{ε_1} depending on ε_1, such that by interpolation and Young's inequality

$$\|[\Delta,\varphi_j]u\|_V \leq \varepsilon\|u\|_{Z_{\nu+3}} + c_{\varepsilon_1}\|u\|_V \leq \|u\|_{\mathcal{Z}_{\nu+3}^\lambda}\left(\varepsilon_1 + \frac{c_{\varepsilon_1}}{\lambda}\right) \quad (4.23)$$

holds for all $j = 0, 1, \ldots, n$ and $u \in \mathcal{Z}_{\nu+3}^\lambda$. Then by the boundedness of the solution operators and (4.23) we end up with

$$\|S_j^F([\Delta,\varphi_j](v_2 - v_1))\|_{\mathcal{Z}_{\nu+3}^\lambda} \leq c_1\|[\Delta,\varphi_j](v_2-v_1)\|_V \leq c_1\left(\varepsilon_1 + \frac{c_{\varepsilon_1}}{\lambda}\right)\|v_2 - v_1\|_{\mathcal{Z}_{\nu+3}^\lambda}. \quad (4.24)$$

Facing the second sum from (4.22) we get

$$\sum_{j=n_1+1}^{n} \left\| S_j^B \begin{pmatrix} \Theta_j^{-1}[\Delta,\varphi_j](v_2 - v_1) \\ \Theta_j^{-1}\Gamma_y[\mathcal{D},\varphi_j](v_1 - v_2) \end{pmatrix} \right\|_{\mathcal{Z}_{\nu+3}^\lambda} \\ \leq c_2 \sum_{j=n_1+1}^{n} \|[\Delta,\varphi_j](v_2-v_1)\|_V + C(\lambda) \sum_{j=n_1+1}^{n} \|\Gamma_y[\mathcal{D},\varphi_j](v_1-v_2)\|_{D(F^{\frac{\nu}{2}})},$$

where $C(\lambda) \sim \lambda^{\frac{\nu}{4}}$ as $\lambda \to \infty$, which is due to Proposition 4.6. The first term is fine by (4.24). For the second we may stress that $u \in Z_{\nu+1}$ implies $\Gamma_y u \in D(F^{\frac{\nu}{2}})$. Now we estimate with the aid of Young's inequality

$$\|\Gamma_y[\mathcal{D},\varphi_j](v_1-v_2)\|_{D(F^{\frac{\nu}{2}})} \leq c_3\|v_1 - v_2\|_{Z_{\nu+1}} \\ \leq c_4(\varepsilon_2\|v_1-v_2\|_{Z_{\nu+3}} + c_{\varepsilon_2}\|v_1-v_2\|_V) \quad (4.25) \\ \leq c_4\left(\varepsilon_2 + \frac{c_{\varepsilon_2}}{\lambda}\right)\|v_1-v_2\|_{\mathcal{Z}_{\nu+3}^\lambda}.$$

Hence, we deduced

$$\|\mathcal{G}(v_1) - \mathcal{G}(v_2)\|_{\mathcal{Z}_{\nu+3}^\lambda} \leq c_5\left[(n+1)\left(\varepsilon_1 + \frac{c_{\varepsilon_1}}{\lambda}\right) + (n-n_1)C(\lambda)\left(\varepsilon_2 + \frac{c_{\varepsilon_2}}{\lambda}\right)\right]\|v_1-v_2\|_{\mathcal{Z}_{\nu+3}^\lambda},$$

where the Lipschitz constant can be made arbitrary small by choosing first ε_1 and ε_2 sufficiently small and then selecting λ appropriately large (recall that by Proposition 4.6 $C(\lambda) \sim \lambda^{\frac{\nu}{4}}$ in a neighborhood of infinity and $\nu < 4$ by assumption).

Summarizing we have shown that the solution v of (4.15) belongs to $Z_{\nu+3}$. Lastly, it is due to [112, Theorem 3.1] that the solution w of the auxiliary problem (4.16) in particular belongs to $Z_{\nu+7}$, which immediately results in the fact that the function $u = v + w$ is a solution of the initial problem (4.1) and, moreover, affiliates to the space $Z_{\nu+3}$. This completes the proof of (ii). The proof of the corresponding Dirichlet problem (i) can be obtained by following the arguments of the proof of (ii). Therefore we omit it.

Appendix A

Basic essentials

For $f \in L_1(\mathbb{R}; X)$, the Fourier transform of f is the function $\mathcal{F}f : \mathbb{R} \to X$ defined by

$$(\mathcal{F}f)(s) = \int_{\mathbb{R}} e^{-ist} f(t) \mathrm{d}t.$$

For scalar-valued functions f Plancherel's Theorem can be found in Rudin [95, Theorem 19.2]. Plancherel's Theorem is not true for vector-valued functions, except when the space X is a Hilbert space. The theorem then reads as

Theorem A.1 (Plancherel's Theorem). *Let X be a Hilbert space. Then $\mathcal{F}f \in L_2(\mathbb{R}; X)$ and $\|\mathcal{F}f\|_{L_2(\mathbb{R};X)} = \sqrt{2\pi} \|f\|_{L_2(\mathbb{R};X)}$ for all $f \in L_1(\mathbb{R}; X) \cap L_2(\mathbb{R}; X)$. The restriction of $\mathcal{F}$ to $L_1(\mathbb{R}; X) \cap L_2(\mathbb{R}; X)$ extends to a bounded linear operator $\mathcal{F}$ on $L_2(\mathbb{R}; X)$ and $(2\pi)^{-1/2} \mathcal{F}$ is an unitary operator on the Hilbert space $L_2(\mathbb{R}; X)$. Moreover,*

$$\int_{\mathbb{R}} ((\mathcal{F}f)(t)|g(t))_X \, \mathrm{d}t = \int_{\mathbb{R}} (f(t)|(\mathcal{F}g)(-t))_X \, \mathrm{d}t$$

for all $f, g \in L_2(\mathbb{R}; X)$.

For the proof we refer to Arendt et al. [12, Proof of Theorem 1.8.2]. Let $\mathbb{C}_+ := \{\lambda \in \mathbb{C} : \operatorname{Re}\lambda > 0\}$ and $H_2(\mathbb{C}_+; X)$ be the space of all holomorphic functions $g : \mathbb{C}_+ \to X$ such that

$$\|g\|_{H_2(\mathbb{C}_+;X)}^2 := \sup_{\alpha > 0} \int_{\mathbb{R}} \|g(\alpha + is)\|^2 \mathrm{d}s < \infty.$$

Let as usual denote $\widehat{f}$ the Laplace transform of the function f. For scalar functions, the Paley-Wiener Theorem can be found in Rudin [95, Theorem 19.13]. Again the Paley-Wiener Theorem is not true for vector-valued functions in general, but it is true in the case of a Hilbert space and then reads

Theorem A.2 (Paley-Wiener Theorem). *Let X be a Hilbert space. Then the map $f \mapsto \widehat{f}|_{\mathbb{C}_+}$ is an isometric isomorphism of $L_2(\mathbb{R}_+; X)$ onto $H_2(\mathbb{C}_+; X)$. Moreover, for $f \in L_2(\mathbb{R}_+; X)$,*

$$\widehat{f}(\alpha + is) = \frac{\alpha}{\pi} \int_{\mathbb{R}} \frac{(\mathcal{F}f)(r)}{\alpha^2 + (s-r)^2} dr.$$

As $\alpha \downarrow 0$, $\|\widehat{f}(\alpha + is) - (\mathcal{F}f)(s)\| \to 0$ (s)-a.e. and $\int_{\mathbb{R}} \|\widehat{f}(\alpha + is) - (\mathcal{F}f)(s)\|^2 ds \to 0$.

The proof can be found in Arendt et al. [12, Proof of Theorem 1.8.3].

We are now turning to the Kahane-Khintchine inequality. To circumvent an introduction to the principle of hypocontractive domination (see Kwapień & Woyczyński [67, Section 3.3]) we rephrase the result of [67, Corollary 3.4.1]. It then reads as

Theorem A.3 (Kahane-Khintchine inequality). *Let X be a Banach space and $1 < q < p < \infty$. If ξ is a centered Gaussian random variable then*

$$\left(\mathbb{E} \left\| x + \sqrt{\frac{q-1}{p-1}} y\xi \right\|_X^p \right)^{1/p} \leq \left(\mathbb{E} \|x + y\xi\|_X^q \right)^{1/q}$$

holds for every $x, y \in X$.

In the present thesis we will frequently make use of the case when $x = 0$, to position ourself to employ the Kolmogorov-Čentsov Theorem.

Theorem A.4 (Kolmogorov-Čentsov Theorem). *Suppose that a process $X := \{X(t)\}_{t\in[0,T]}$ on a probability space $(\Omega, \mathscr{F}, \mathbb{P})$ satisfies the condition*

$$\mathbb{E}|X(t) - X(s)|^\alpha \leq C|t-s|^{1+\beta}, \qquad s, t \in [0, T],$$

for some positive constants α, β, and C. Then there exists a continuous modification of X, which is locally Hölder-continuous with exponent ν for every $\nu \in (0, \frac{\beta}{\alpha})$.

A very elementary proof based on Čebyšev's inequality can be found in Karatzas & Shreve [60, Proof of Theorem 2.8]. For a more analytic version of the proof we refer to Da Prato & Zabczyk [27, Proof of Theorem 3.3].

List of Figures

1   Idealized regions for the values of $\mathrm{Var}[X(t)]$, where X is centered and due to (ϕ) and (ϕ$_0$) with $\theta = 0$ (left) and $\theta > 0$ (right), respectively.   9

2.1 A sample path of a fractional Brownian motion with Hurst parameter $H = 0.2$. 75

2.2 A sample path of a fractional Brownian motion with Hurst parameter $H = 0.5$. 75

2.3 A sample path of fractional Brownian motion with Hurst parameter $H = 0.9$. 75

2.4 Sample of $1/f^0$-noise. 77

2.5 Sample of $1/f^{1/2}$-noise. 77

2.6 Sample of $1/f^1$-noise. 78

2.7 The actual values (dashed) and the predicted region of the second moments $\mathbb{E}[X(t)]^2$ with $X = \mathcal{RB}_\alpha^\beta$, where $\alpha = 0.1$ and $\beta = 0.6$ (left) resp. $\alpha = 0.89$ and $\beta = 0.6$ (right). 82

2.8 The actual values (dashed) and the predicted region of the second moments $\mathbb{E}[X(t)]^2$ with $X = \mathcal{RB}_\alpha^\beta$, where $\alpha = 0.05$ and $\beta = 1.2$ (left) resp. $\alpha = 0.29$ and $\beta = 1.2$ (right). 82

2.9 A random path of a fractional Riesz-Bessel motion (left) and corresponding noise profile (right) with parameters $\beta = 0.7$ and $\alpha = 0.1$. 84

2.10 A random path of a fractional Riesz-Bessel motion (left) and corresponding noise profile (right) with parameters $\beta = 1.05$ and $\alpha = 0.1$. 84

2.11 A random path of a fractional Riesz-Bessel motion (left) and corresponding noise profile (right) with parameters $\beta = 1.05$ and $\alpha = 0.44$. 84

2.12 A random path of a fractional Riesz-Bessel motion (left) and corresponding noise profile (right) with parameters $\beta = 1.05$ and $\alpha = 300$. 85

Bibliography

[1] E. Alòs, J. A. León, and D. Nualart, *Stochastic Stratonovich calculus fBm for fractional Brownian motion with Hurst parameter less than* $1/2$, Taiwanese J. Math. **5** (2001), no. 3, 609–632.

[2] E. Alòs, O. Mazet, and D. Nualart, *Stochastic calculus with respect to fractional Brownian motion with Hurst parameter lesser than* $\frac{1}{2}$, Stochastic Process. Appl. **86** (2000), no. 1, 121–139.

[3] E. Alòs and D. Nualart, *Stochastic integration with respect to the fractional Brownian motion*, Stoch. Stoch. Rep. **75** (2003), no. 3, 129–152.

[4] H. Amann, *Linear and quasilinear parabolic problems. Vol. I*, Monographs in Mathematics, vol. 89, Birkhäuser Boston Inc., Boston, MA, 1995, Abstract linear theory.

[5] J. M. Angulo, M. D. Ruiz-Medina, V. V. Anh, and W. Grecksch, *Fractional diffusion and fractional heat equation*, Adv. in Appl. Probab. **32** (2000), no. 4, 1077–1099.

[6] V. V. Anh, J. M. Angulo, and M. D. Ruiz-Medina, *Possible long-range dependence in fractional random fields*, J. Statist. Plann. Inference **80** (1999), no. 1-2, 95–110.

[7] V. V. Anh and W. Grecksch, *A fractional stochastic evolution equation driven by fractional Brownian motion*, Monte Carlo Methods Appl. **9** (2003), no. 3, 189–199.

[8] V. V. Anh, C. C. Heyde, and Q. Tieng, *Stochastic models for fractal processes*, J. Statist. Plann. Inference **80** (1999), no. 1-2, 123–135.

[9] V. V. Anh and C. N. Nguyen, *Stochastic analysis of fractional Riesz-Bessel motion*, Random Oper. Stochastic Equations **8** (2000), no. 2, 105–126.

[10] _____, *Semimartingale representation of fractional Riesz-Bessel motion*, Finance Stoch. **5** (2001), no. 1, 83–101.

[11] D. Applebaum, *Lévy processes and stochastic calculus*, Cambridge Studies in Advanced Mathematics, vol. 93, Cambridge University Press, Cambridge, 2004.

[12] W. Arendt, C. J. K. Batty, M. Hieber, and F. Neubrander, *Vector-valued Laplace transforms and Cauchy problems*, Monographs in Mathematics, vol. 96, Birkhäuser Verlag, Basel, 2001.

[13] C. Bender, *An Itô formula for generalized functionals of a fractional Brownian motion with arbitrary Hurst parameter*, Stochastic Process. Appl. **104** (2003), no. 1, 81–106.

[14] C. Bender, T. Sottinen, and E. Valkeila, *Arbitrage with fractional Brownian motion?*, Theory Stoch. Process. **13** (2007), no. 1-2, 23–34.

[15] J. Beran, *Statistics for long-memory processes*, Monographs on Statistics and Applied Probability, vol. 61, Chapman and Hall, New York, 1994.

[16] R. Bhansali, M. P. Holland, and P. S. Kokoszka, *Intermittency, long-memory and financial returns*, Long memory in economics, Springer, Berlin, 2007, pp. 39–68.

[17] F. Biagini, Y. Hu, B. Øksendal, and T. Zhang, *Stochastic calculus for fractional Brownian motion and applications*, Probability and its Applications (New York), Springer-Verlag London Ltd., London, 2008.

[18] F. Biagini and B. Øksendal, *Forward integrals and an Itô formula for fractional Brownian motion*, Infin. Dimens. Anal. Quantum Probab. Relat. Top. **11** (2008), no. 2, 157–177.

[19] F. Biagini, B. Øksendal, A. Sulem, and N. Wallner, *An introduction to white-noise theory and Malliavin calculus for fractional Brownian motion*, Proc. R. Soc. Lond. Ser. A Math. Phys. Eng. Sci. **460** (2004), no. 2041, 347–372, Stochastic analysis with applications to mathematical finance. MR MR2052267 (2005a:60107)

[20] S. Bonaccorsi, *Volterra equations perturbed by a Gaussian noise*, Seminar on Stochastic Analysis, Random Fields and Applications V, Progr. Probab., vol. 59, Birkhäuser, Basel, 2008, pp. 37–55.

[21] P. L. Butzer and U. Westphal, *An introduction to fractional calculus*, Applications of fractional calculus in physics, World Sci. Publ., River Edge, NJ, 2000, pp. 1–85.

[22] M. Caputo, *Linear models of dissipation whose Q is almost frequency independent. II*, Fract. Calc. Appl. Anal. **11** (2008), no. 1, 4–14, Reprinted from Geophys. J. R. Astr. Soc. **13** (1967), no. 5, 529–539.

[23] Ph. Carmona, L. Coutin, and G. Montseny, *Stochastic integration with respect to fractional Brownian motion*, Ann. Inst. H. Poincaré Probab. Statist. **39** (2003), no. 1, 27–68.

[24] P. Cheridito, *Arbitrage in fractional Brownian motion models*, Finance Stoch. **7** (2003), no. 4, 533–553.

[25] Ph. Clément, G. Da Prato, and J. Prüss, *White noise perturbation of the equations of linear parabolic viscoelasticity*, Rend. Istit. Mat. Univ. Trieste **29** (1997), no. 1-2, 207–220 (1998).

[26] L. Coutin, D. Nualart, and C. A. Tudor, *Tanaka formula for the fractional Brownian motion*, Stochastic Process. Appl. **94** (2001), no. 2, 301–315.

[27] G. Da Prato and J. Zabczyk, *Stochastic equations in infinite dimensions*, Encyclopedia of Mathematics and its Applications, vol. 44, Cambridge University Press, Cambridge, 1992.

[28] W. Dai and C. C. Heyde, *Itô's formula with respect to fractional Brownian motion and its application*, J. Appl. Math. Stochastic Anal. **9** (1996), no. 4, 439–448.

[29] A. B. Davis, A. L. Marshak, W. J. Wiscombe, and R. F. Cahalan, *Multifractal characterizations of intermittency in nonstationary geophysical signals and fields: a model-based perspective on ergodicity issues illustrated with cloud data*, Current topics in nonstationary analysis (San Diego, CA, 1995), World Sci. Publ., River Edge, NJ, 1996, pp. 97–158.

[30] L. Decreusefond, *A Skohorod-Stratonovitch integral for the fractional Brownian motion*, Stochastic analysis and related topics, VII (Kusadasi, 1998), Progr. Probab., vol. 48, Birkhäuser Boston, Boston, MA, 2001, pp. 177–198.

[31] _____, *Stochastic integration with respect to fractional Brownian motion*, Theory and applications of long-range dependence, Birkhäuser Boston, Boston, MA, 2003, pp. 203–226.

[32] L. Decreusefond and A. S. Üstünel, *Fractional Brownian motion: theory and applications*, Systèmes différentiels fractionnaires (Paris, 1998), ESAIM Proc., vol. 5, Soc. Math. Appl. Indust., Paris, 1998, pp. 75–86.

[33] _____, *Stochastic analysis of the fractional Brownian motion*, Potential Anal. **10** (1999), no. 2, 177–214.

[34] R. Denk, M. Hieber, and J. Prüss, *$\mathcal{R}$-boundedness, Fourier multipliers and problems of elliptic and parabolic type*, Mem. Amer. Math. Soc. **166** (2003), no. 788, viii+114.

[35] T. E. Duncan, Y. Hu, and B. Pasik-Duncan, *Stochastic calculus for fractional Brownian motion. I. Theory*, SIAM J. Control Optim. **38** (2000), no. 2, 582–612 (electronic).

[36] T. E. Duncan, J. Jakubowski, and B. Pasik-Duncan, *Stochastic integration for fractional Brownian motion in a Hilbert space*, Stoch. Dyn. **6** (2006), no. 1, 53–75.

[37] T. E. Duncan, B. Pasik-Duncan, and B. Maslowski, *Fractional Brownian motion and stochastic equations in Hilbert spaces*, Stoch. Dyn. **2** (2002), no. 2, 225–250.

[38] N. Dunford and J. T. Schwartz, *Linear operators. Part I*, Wiley Classics Library, John Wiley & Sons Inc., New York, 1988, General theory, With the assistance of William G. Bade and Robert G. Bartle, Reprint of the 1958 original, A Wiley-Interscience Publication.

[39] _____, *Linear operators. Part II*, Wiley Classics Library, John Wiley & Sons Inc., New York, 1988, Spectral theory. Selfadjoint operators in Hilbert space, With the assistance of William G. Bade and Robert G. Bartle, Reprint of the 1963 original, A Wiley-Interscience Publication.

[40] M. M. Džrbašjan and A. B. Nersesjan, *Fractional derivatives and the Cauchy problem for differential equations of fractional order*, Izv. Akad. Nauk Armjan. SSR Ser. Mat. **3** (1968), no. 1, 3–29.

[41] R. J. Elliott and J. van der Hoek, *Fractional Brownian motion and financial modelling*, Mathematical finance (Konstanz, 2000), Trends Math., Birkhäuser, Basel, 2001, pp. 140–151.

[42] _____, *Itô formulas for fractional Brownian motion*, Advances in mathematical finance, Appl. Numer. Harmon. Anal., Birkhäuser Boston, Boston, MA, 2007, pp. 59–81.

[43] K. Falconer, *Fractal geometry*, second ed., John Wiley & Sons Inc., Hoboken, NJ, 2003, Mathematical foundations and applications.

[44] U. Frisch, *Turbulence*, Cambridge University Press, Cambridge, 1995, The legacy of A. N. Kolmogorov.

[45] I. M. Gelfand and N. Ya. Vilenkin, *Generalized functions. Vol. 4*, Academic Press [Harcourt Brace Jovanovich Publishers], New York, 1964 [1977], Applications of harmonic analysis, Translated from the Russian by Amiel Feinstein.

[46] W. G. Glöckle and T. F. Nonnenmacher, *Fox function representation of non-Debye relaxation processes*, J. Statist. Phys. **71** (1993), no. 3-4, 741–757.

[47] R. Gorenflo and F. Mainardi, *Fractional calculus: integral and differential equations of fractional order*, Fractals and fractional calculus in continuum mechanics (Udine, 1996), CISM Courses and Lectures, vol. 378, Springer, Vienna, 1997, pp. 223–276.

[48] _____, *Fractional relaxation of distributed order*, Complexus mundi, World Sci. Publ., Hackensack, NJ, 2006, pp. 33–42.

[49] M. Gradinaru, I. Nourdin, F. Russo, and P. Vallois, *m-order integrals and generalized Itô's formula: the case of a fractional Brownian motion with any Hurst index*, Ann. Inst. H. Poincaré Probab. Statist. **41** (2005), no. 4, 781–806.

[50] P. Guasoni, *No arbitrage under transaction costs, with fractional Brownian motion and beyond*, Math. Finance **16** (2006), no. 3, 569–582.

[51] J. A. Gubner, *Theorems and fallacies in the theory of long-range-dependent processes*, IEEE Trans. Inform. Theory **51** (2005), no. 3, 1234–1239.

[52] G. H. Hardy, J. E. Littlewood, and G. Pólya, *Inequalities*, Cambridge Mathematical Library, Cambridge University Press, Cambridge, 1988, Reprint of the 1952 edition.

[53] R. Hilfer, *Fractional time evolution*, Applications of fractional calculus in physics, World Sci. Publ., River Edge, NJ, 2000, pp. 87–130.

[54] Y. Hu, *Option pricing in a market where the volatility is driven by fractional Brownian motions*, Recent developments in mathematical finance (Shanghai, 2001), World Sci. Publ., River Edge, NJ, 2002, pp. 49–59.

[55] H. E. Hurst, R. P. Black, and Y. M. Sinaika, *Long term storage. An experimental study*, Constable, London, 1965.

[56] K. Itô, *Stochastic integral*, Proc. Imp. Acad. Tokyo **20** (1944), 519–524.

[57] M. Jolis, *On the Wiener integral with respect to the fractional Brownian motion on an interval*, J. Math. Anal. Appl. **330** (2007), no. 2, 1115–1127.

[58] G. Jumarie, *Merton's model of optimal portfolio in a Black-Scholes market driven by a fractional Brownian motion with short-range dependence*, Insurance Math. Econom. **37** (2005), no. 3, 585–598.

[59] N. J. Kalton and L. Weis, *The H^∞-calculus and sums of closed operators*, Math. Ann. **321** (2001), no. 2, 319–345.

[60] I. Karatzas and S. E. Shreve, *Brownian motion and stochastic calculus*, second ed., Graduate Texts in Mathematics, vol. 113, Springer-Verlag, New York, 1991.

[61] M. Kijima and T. Suzuki, *The pricing of options with stochastic boundaries in a Gaussian economy*, J. Oper. Res. Soc. Japan **50** (2007), no. 2, 137–150.

[62] V. L. Kobelev, O. L. Kobeleva, Ya. L. Kobelev, and L. Ya. Kobelev, *Diffusion through a fractal surface*, Dokl. Akad. Nauk **355** (1997), no. 3, 326–327.

[63] V. L. Kobelev, E. P. Romanov, Ya. L. Kobelev, and L. Ya. Kobelev, *Non-Debye relaxation and diffusion in a fractal space*, Dokl. Akad. Nauk **361** (1998), no. 6, 755–758.

[64] Ya. L. Kobelev, L. Ya. Kobelev, V. L. Kobelev, and E. P. Romanov, *Diffusion in fractal media on the basis of the Klimontovich kinetic equation in a fractal space*, Dokl. Akad. Nauk **385** (2002), no. 5, 612–614.

[65] A. N. Kolmogorov, *Curves in Hilbert space which are invariant with respect to a one-parameter group of motions*, Dokl. Akad. Nauk SSSR **26** (1940), no. 1, 115–118.

[66] Y. V. Krvavych and Y. S. Mishura, *The stochastic Fubini theorem for integrals containing random integrand and fractional Brownian motion as integrator*, Theory Stoch. Process. **6** (2000), no. 1-2, 79–89.

[67] S. Kwapień and W. A. Woyczyński, *Random series and stochastic integrals: single and multiple*, Probability and its Applications, Birkhäuser Boston Inc., Boston, MA, 1992.

[68] H. Lakhel, Y. Ouknine, and C. A. Tudor, *Besov regularity for the indefinite Skorohod integral with respect to the fractional Brownian motion: the singular case*, Stoch. Stoch. Rep. **74** (2002), no. 3-4, 597–615.

[69] S. J. Lin, *Stochastic analysis of fractional Brownian motions*, Stochastics Stochastics Rep. **55** (1995), no. 1-2, 121–140.

[70] S. Y. Liu and X. Q. Yang, *Pricing of European contingent claim in fractional Brownian motion environment*, Chinese J. Appl. Probab. Statist. **20** (2004), no. 4, 429–434.

[71] _____, *Pricing of compound option in a fractional Brownian motion environment*, Gongcheng Shuxue Xuebao **23** (2006), no. 1, 153–157.

[72] G. Lumer and R. S. Phillips, *Dissipative operators in a Banach space*, Pacific J. Math. **11** (1961), 679–698.

[73] F. Mainardi and R. Gorenflo, *Time-fractional derivatives in relaxation processes: a tutorial survey*, Fract. Calc. Appl. Anal. **10** (2007), no. 3, 269–308.

[74] F. Mainardi and P. Paradisi, *Fractional diffusive waves*, J. Comput. Acoust. **9** (2001), no. 4, 1417–1436.

[75] B. B. Mandelbrot and R. L. Hudson, *The (mis)behavior of markets*, Basic Books, New York, 2004, A fractal view of risk, ruin, and reward.

[76] B. B. Mandelbrot and J. W. Van Ness, *Fractional Brownian motions, fractional noises and applications*, SIAM Rev. **10** (1968), 422–437.

[77] K. S. Miller and B. Ross, *An introduction to the fractional calculus and fractional differential equations*, A Wiley-Interscience Publication, John Wiley & Sons Inc., New York, 1993.

[78] S. Monniaux and J. Prüss, *A theorem of the Dore-Venni type for noncommuting operators*, Trans. Amer. Math. Soc. **349** (1997), no. 12, 4787–4814.

[79] C. Necula, *Option pricing in a fractional Brownian motion environment*, Math. Rep. (Bucur.) **6(56)** (2004), no. 3, 259–273.

[80] I. Norros, E. Valkeila, and J. Virtamo, *An elementary approach to a Girsanov formula and other analytical results on fractional Brownian motions*, Bernoulli **5** (1999), no. 4, 571–587.

[81] D. Nualart, *Stochastic integration with respect to fractional Brownian motion and applications*, Stochastic models (Mexico City, 2002), Contemp. Math., vol. 336, Amer. Math. Soc., Providence, RI, 2003, pp. 3–39. MR MR2037156 (2004m:60119)

[82] _____, *Stochastic calculus with respect to fractional Brownian motion*, Ann. Fac. Sci. Toulouse Math. (6) **15** (2006), no. 1, 63–78.

[83] B. Øksendal, *Fractional Brownian motion in finance*, Stochastic economic dynamics, Cph. Bus. Sch. Press, Frederiksberg, 2007, pp. 11–56.

[84] K. B. Oldham and J. Spanier, *The fractional calculus*, Academic Press [A subsidiary of Harcourt Brace Jovanovich, Publishers], New York-London, 1974,

Theory and applications of differentiation and integration to arbitrary order, With an annotated chronological bibliography by Bertram Ross, Mathematics in Science and Engineering, Vol. 111.

[85] E. E. Peters, *Fractal Market Analysis*, Wiley, New York, 1994, Applying Chaos Theory to Investment and Economics.

[86] V. Pipiras and M. S. Taqqu, *Integration questions related to fractional Brownian motion*, Probab. Theory Related Fields **118** (2000), no. 2, 251–291.

[87] I. Podlubny, *Fractional differential equations*, Mathematics in Science and Engineering, vol. 198, Academic Press Inc., San Diego, CA, 1999, An introduction to fractional derivatives, fractional differential equations, to methods of their solution and some of their applications.

[88] J. Prüss, *Evolutionary integral equations and applications*, Monographs in Mathematics, vol. 87, Birkhäuser Verlag, Basel, 1993.

[89] _____, *Maximal regularity for abstract parabolic problems with inhomogeneous boundary data in L_p-spaces*, Proceedings of EQUADIFF, 10 (Prague, 2001), vol. 127, 2002, pp. 311–327.

[90] _____, *Maximal regularity for evolution equations in L_p-spaces*, Conf. Semin. Mat. Univ. Bari (2002), no. 285, 1–39 (2003).

[91] J. Prüss, S. Sperlich, and M. Wilke, *An analysis of Asian options*, Functional analysis and evolution equations, Birkhäuser, Basel, 2008, pp. 547–559.

[92] L. C. G. Rogers, *Equivalent martingale measures and no-arbitrage*, Stochastics Stochastics Rep. **51** (1994), no. 1-2, 41–49.

[93] A. Rößler, M. Seaïd, and M. Zahri, *Method of lines for stochastic boundary-value problems with additive noise*, Appl. Math. Comput. **199** (2008), no. 1, 301–314.

[94] Y. A. Rozanov and F. Sanso, *On stochastic boundary value problems for harmonic functions in Banach spaces*, Teor. Veroyatnost. i Primenen. **44** (1999), no. 1, 123–128.

[95] W. Rudin, *Real and complex analysis*, third ed., McGraw-Hill Book Co., New York, 1987.

[96] M. D. Ruiz-Medina, J. M. Angulo, and V. V. Anh, *Fractional generalized random fields on bounded domains*, Stochastic Anal. Appl. **21** (2003), no. 2, 465–492.

[97] Th. Runst and W. Sickel, *Sobolev spaces of fractional order, Nemytskij operators, and nonlinear partial differential equations*, de Gruyter Series in Nonlinear Analysis and Applications, vol. 3, Walter de Gruyter & Co., Berlin, 1996.

[98] A. I. Saichev and G. M. Zaslavsky, *Fractional kinetic equations: solutions and applications*, Chaos **7** (1997), no. 4, 753–764.

[99] S. G. Samko, A. A. Kilbas, and O. I. Marichev, *Fractional integrals and derivatives*, Gordon and Breach Science Publishers, Yverdon, 1993, Theory and applications, Edited and with a foreword by S. M. Nikol'skiĭ, Translated from the 1987 Russian original, Revised by the authors.

[100] E. Scalas, R. Gorenflo, and F. Mainardi, *Fractional calculus and continuous-time finance*, Phys. A **284** (2000), no. 1-4, 376–384.

[101] A. N. Shiryaev, *Essentials of stochastic finance*, Advanced Series on Statistical Science & Applied Probability, vol. 3, World Scientific Publishing Co. Inc., River Edge, NJ, 1999, Facts, models, theory, Translated from the Russian manuscript by N. Kruzhilin.

[102] A. V. Skorohod, *On a generalization of the stochastic integral*, Teor. Verojatnost. i Primenen. **20** (1975), no. 2, 223–238.

[103] S. Sperlich, *On parabolic Volterra equations disturbed by fractional Brownian motions*, Stoch. Anal. Appl. **27** (2009), no. 1, 74–94.

[104] S. Sperlich and M. Wilke, *Fractional white noise perturbations of parabolic Volterra equations*, to appear in J. Appl. Anal. (2009), Preprint: http://www.mathematik.uni-halle.de/reports/sources/2007/07-10report.pdf, 2007.

[105] R. L. Stratonovič, *A new form of representing stochastic integrals and equations*, Vestnik Moskov. Univ. Ser. I Mat. Meh. **1964** (1964), no. 1, 3–12.

[106] H. Triebel, *Theory of function spaces*, Monographs in Mathematics, vol. 78, Birkhäuser Verlag, Basel, 1983.

[107] C. Tudor, *On the Wiener integral with respect to the fractional Brownian motion*, Bol. Soc. Mat. Mexicana (3) **8** (2002), no. 1, 97–106.

[108] C. A. Tudor, *Itô formula for the infinite-dimensional fractional Brownian motion*, J. Math. Kyoto Univ. **45** (2005), no. 3, 531–546.

[109] J. von Neumann and I. J. Schoenberg, *Fourier integrals and metric geometry*, Trans. Amer. Math. Soc. **50** (1941), 226–251.

[110] A. M. Yaglom, *Correlation theory of stationary and related random functions. Vol. I*, Springer Series in Statistics, Springer-Verlag, New York, 1987, Basic results.

[111] R. Zacher, *Quasilinear parabolic problems with nonlinear boundary conditions*, Ph.D. thesis, Martin-Luther-Universität Halle-Wittenberg, 2003, http://sundoc.bibliothek.uni-halle.de/diss-online/03/03H058/prom.pdf.

[112] _____, *Maximal regularity of type L_p for abstract parabolic Volterra equations*, J. Evol. Equ. **5** (2005), no. 1, 79–103.

[113] _____, *Quasilinear parabolic integro-differential equations with nonlinear boundary conditions*, Differential Integral Equations **19** (2006), no. 10, 1129–1156.

i want morebooks!

Buy your books fast and straightforward online - at one of world's fastest growing online book stores! Environmentally sound due to Print-on-Demand technologies.

Buy your books online at
www.get-morebooks.com

Kaufen Sie Ihre Bücher schnell und unkompliziert online – auf einer der am schnellsten wachsenden Buchhandelsplattformen weltweit! Dank Print-On-Demand umwelt- und ressourcenschonend produziert.

Bücher schneller online kaufen
www.morebooks.de

VDM Verlagsservicegesellschaft mbH
Heinrich-Böcking-Str. 6-8   Telefon: +49 681 3720 174   info@vdm-vsg.de
D - 66121 Saarbrücken   Telefax: +49 681 3720 1749   www.vdm-vsg.de

Printed by Books on Demand GmbH, Norderstedt / Germany